prehistoric
past
revealed

Douglas Palmer

prehistoric
past
revealed

The four billion year
history of life on Earth

University of California Press
Berkeley   Los Angeles   London

University of California Press
Berkeley and Los Angeles, California

Published by arrangement with Mitchell Beazley,
an imprint of Octopus Publishing Group Limited

First published in 2003
© Douglas Palmer

ISBN 0-520-24105-3

Cataloguing-in-Publication data is on file
with the Library of Congress

10 9 8 7 6 5 4 3 2 1

Page 2: Fossil leaf from Messel Pit, Germany
page 5: Caudipteryx, small Cretaceous bird-like dinosaur

Set in DINEngschrift and Bliss

Printed and bound in China

Contents

introduction

The aim of *Prehistoric Past Revealed* is to explore the geological history of the Earth and its inhabitants from the present day with its more familiar environments and life, and work back progressively through time into the less well-known depths of the Earth's geological past.

Although this narrative goes against the flow of time and evolution, it follows the same path as the discovery of Earth history . Generally, the youngest and most recent sediments, rocks, and layers of strata are either at or close to the present land surface. These recent "pages" of the Earth story tend to be less disturbed by geological processes such as faulting, folding, or metamorphism than the older ones, simply because they have not been around for long enough to suffer such processes. At least this is the situation in northern Europe, where so much of the early exploration of Earth history took place. Elsewhere, in places such as Mediterranean Europe or California, this generalization does not hold. Young strata are highly deformed by recent Earth movements and continuing mountain building. But in both types of geological environment, fossils found in young strata are more closely related to living forms than those found in more ancient strata. The further back in time you go, the greater the proportion of the fossil biota that belongs to extinct and less familiar groups of animals and plants.

By accident of geological history, northern Europe, and especially the British Isles, contain within a very small geographical area a remarkably good sample of the last 1,000 million years or so of Earth history. Across Britain from south to north, younger to older successive strata are laid out like the pages of a book waiting to be leafed through. However, the book and its narrative was read backwards, from the most recent ie the present, towards the beginning, the most ancient. And it soon became evident that the closer to the beginning you went, the more pages and sometimes whole chapters were missing. Furthermore, the earlier pages are often torn and crumpled so that it is difficult to decipher the text.

Then it was realized that there was not just one copy of this book, but that each continental region had its own edition, a bit like the different gospels of the New Testament. In order to get the whole story it was necessary to try to match the different versions. This task is still not complete, but is being worked upon all the time.

When the geological vastness of the great continents of the Americas, Africa, Asia, Australia etc gradually became better known, it was also realized that the European version of the story was very much a miniturized version. Many of the most important geological processes and products, from river systems to mountain ranges, are normally on a scale which are an order of magnitude larger than those represented by so many European rocks. The Thames or the Rhine are small compared with the Amazon, Indus, or Mississippi. Likewise the ancient Paleozoic "mountains" of Scotland pale by comparison with the Himalayas and Andes or Western Cordillera of the Americas.

Perhaps the greatest shift in understanding of the geological past came with new investigative techniques that were driven forwards by the necessities of the Industrial Revolution. For the first time geology became a modern profession and academic subject of investigation with the establishment of national surveys and university departments. But it was the dire necessities of World War II, which drove the new techniques forward and were to bring a revolution in geological understanding. The accurate dating of igneous rocks using radioisotopes took a long time to become a reliable everyday technique. But by the 1950s, radiometric dating promoted a great leap forward, as did the use of earthquake waves to explore the inaccessible depths of the Earth, revealing its layered structure. The measurement of rock magnetism, combined with the detailed mapping of the ocean floor, all provided essential ingredients which fuelled the plate tectonic revolution.

These developments caused a radical reappraisal of the old methodologies and paradigms about the past, especially concerning the early history of the Earth. It was realized that there are many geological processes which cannot be easily assessed by what happens within a daily, annual, or even decadal scale. Many important events are very rare and others happen so slowly that we are only now recognizing their significance. Since astronomy put the Earth and its formation in the context of the development of the Solar System and beyond that in a galactic context, it has been realized that conditions and environments in the early history of the Earth were very different from those of the present. The present is not always a key to the past.

How to use the timelines in this book

Geological time has a very complicated structure, with many subdivisions and sub-subdivisions. This has evolved as geologists have discovered more about the age of the Earth and major events in its history.

In this book, we are using four main levels, or divisions of time. They are the eon (the biggest), followed by the era, then the period, and finally the epoch (the smallest); that is, several epochs go together to form a period, several periods form an era, and several eras comprise an eon. We have devised timelines to take you through these time spans in easy stages.

Each chapter has a double timeline on its opening spread. The top part shows the eons, and the eras within those eons, of all time in Earth's history. This top section is the same on every chapter's timeline. Part of this top line is then expanded in the lower section to show the tranche of time being discussed in that particular chapter. This lower line therefore shows these selected eras, and the periods within them.

Each chapter is comprised of several separate topics, listed in the Contents. Where these topics cover only a section of the time covered by the chapter overall, they are given their own single timeline. This shows the period or periods (which you will have already seen on the bottom section of the chapter timeline) relevant to that topic, and the epochs within them.

Chapter 9 is slightly different, since it talks about Earth's future. Thus the bottom section of this chapter's timeline travels millions of years into the future, where we do not know what any new eras or periods might be called.

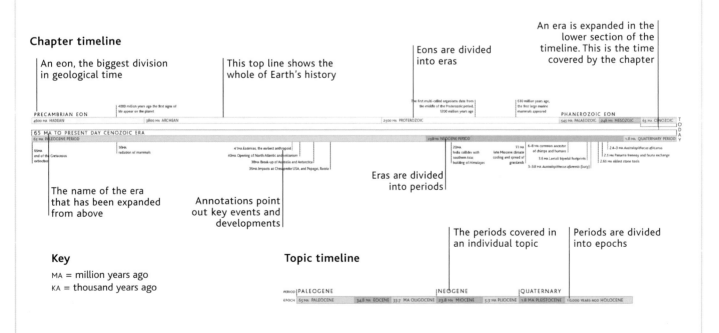

1
How the present reveals the past

Nowadays it seems perfectly reasonable to study present phenomena in order to understand and interpret the past. Observing and measuring volcanoes, earthquakes, glaciation, climate change, weathering, erosion, and sedimentation, and studying their products is an obvious precursor to looking at rocks in order to work out how they were made. Similarly, the study of the biology of living organisms and their relationship with their living environments, known as ecology, can show how fossils fit into the natural world of the past.

However, this approach only became central to the practice of modern geology in the early 1800s, when taken up by Charles Lyell (1797–1875) in his highly influential book *Principles of Geology* (1830–3). The method was pioneered by the Scottish geologist James Hutton (1726–97), and the French anatomist Georges Cuvier (1769–1832) who took similar approaches but came from very different intellectual and cultural backgrounds, with Hutton a product of the Scottish Enlightenment and Cuvier a survivor of the French Revolution who espoused catastrophism.

PRECAMBRIAN EON		3800 MA first evidence of chemical life on the planet		
ERA	4600 MA HADEAN		3800 MA ARCHEAN	

PALEOZOIC ERA					
PERIOD	545 MA CAMBRIAN	495 MA ORDOVICIAN	443 MA SILURIAN 417 MA DEVONIAN	354 MA CARBONIFEROUS	290
	515 MA Jawless fish Jawed fish appear		Upright-growing land plants appear	Forests develop; the first four-legged vertebrates appear	The appearance of flying insects and amphibians Reptiles appear during the late Carboniferous period
	525 MA Radiation of invertebrates				
	545 MA Many small shelly fossils				

Cuvier studied the anatomy of different animal species alive today to see how body structures are modified for particular modes of life. From this he was able to work out the form and habits of extinct organisms even from only partial fossil remains. However, this method had its limitations; when Cuvier first saw the teeth of an *Iguanodon* dinosaur he thought they were rhinoceros teeth. Unsurprisingly, he could not predict the exact form of a completely extinct group of animals just from their teeth; nobody could.

Similarly, many rocks were difficult to interpret because the processes which form them, such as the deep intrusion of granite could not be directly observed. Also many geological processes cannot be observed in real time. Even more problematic were indications that events and environmental conditions were very different in the past. We now realize that there are events such as large scale impacts and glaciations which have not happened recently. Furthermore, the scale and frequency of many events are only now being fully appreciated.

Prehistorical records
Discovered in 1991, about 130m (426.5ft) below present sealevel off France's Mediterranean coast, the Cosquer cave contains an amazing array of Palaeolithic art, dated at 27,000 and 18,500 years ago. Cave art, such as this wonderfully observed horse provides evidence about Ice Age animal anatomy, not available from their fossil remains.

1200 MA The first multi-celled organisms date from the middle of the Proterozoic period

610 MA The first large marine animals appear

PHANEROZOIC EON

| 2500 MA PROTEROZOIC | | 545 MA PALEOZOIC | 248 MA MESOZOIC | 65 MA CENOZOIC | TODAY |

MESOZOIC ERA   CENOZOIC ERA

| ...MIAN | 248 MA TRIASSIC | 205 MA JURASSIC | 142 MA CRETACEOUS | 65 MA PALEOGENE | 23.8 MA NEOGENE | 1.8 MA QUATERNARY |

The first dinosaurs and early mammals appear

Birds and flowering plants appear in the late Jurassic period

Primates and songbirds appear

7 MA First hominids appear on Earth

The diversity of life today

Exploration of the Earth over the past few hundred years has gradually revealed an extraordinary abundance and diversity of life: from microbes living hundreds of metres below ground to coral reefs so large that they are visible from space. It also turns out that this diversity is far from equally distributed around the globe. Important diversity "hotspots" have been identified, especially in the tropics with their rainforests and coral seas. The same historical period has seen the exploration of the rocks and geological formations of the Earth. As successively deeper layers of strata were investigated, fossils were uncovered, revealing that the Earth and its diverse life have an extraordinary history, one that extends far back into the depths of geological time. So what does that history reveal about the origins of our planet and its multifarious inhabitants ranging from microbes to humankind?

The scientific understanding developed over the past few hundred years has revolutionized our view of life, its history and its origins. In the middle of the 18th century, only a few thousand different kinds of living organisms were known. From originally thinking that the Earth and its life were created over a short period of time in the not too distant past, scientists now have compelling evidence that the Earth is some 4600 million years old and that life probably originated as long as 4000 million years ago. Over this remarkably prolonged time span, life has evolved, increased, and diversified hugely, only to be dramatically and drastically cut back and yet recover. Perhaps surprisingly, it is still unclear just how abundant life on Earth is. And many of the details of life's history remain buried deep within the strata of the Earth awaiting the attention of future generations of scientific investigators.

Today, life occupies almost all niches – from deep below ground to the permanently frozen peaks of high mountains, from the depths of the oceans to high in the atmosphere. But this has not always been the case. At its very beginnings, life was confined to water for what was actually a very long period, and the colonization of land and air took some time.

Life's abundance and diversity today

J. B. S. Haldane (1892–1964) once remarked that, if there were a creator, then he or she "must have been inordinately fond of beetles". Despite all the efforts of scientists over the past few centuries, only about 1.7 million of the species of organism alive today have been properly described. Current estimates for the grand total of species of living organisms range between a very conservative eight million and a rather optimistic 100 million, with 30 million seeming the most reasonable estimate. But our understanding of the microbial world of bacteria and more primitive organisms is only just beginning. We may yet be underestimating the total by an order of magnitude and 100 million may indeed be more accurate. Perhaps the arthropods will yet be displaced from their dominant position as the biggest and most diverse group of organisms on Earth.

Both Alfred Russel Wallace (1823–1913) and Charles Darwin (1809–82) were profoundly impressed by the profligacy of life. They were also fascinated by the realization that, if all the offspring of almost any organism were to survive, within a few generations they would vastly outnumber all other creatures around them, dominate their living environment, and run out of

Arthropods

Of all the living organisms known today, a remarkable number are arthropods, especially insects. Arthropods have segmented bodies, paired and jointed limbs, and a toughened organic "skin" known as a cuticle. It has recently been estimated that there are between four and six million arthropod species alive today, among which perhaps half a million are beetles. By comparison, there are a mere 9000 species of bird, which in turn outnumber the 4000 or so mammal species. While these comparisons are quite interesting, however, they are not particularly useful because we are not comparing similar classificatory categories. Only those arthropods with mineralized skeletons such as the extinct trilobites, and crabs are well preserved in the fossil record hence our knowledge of ancient arthropod diversity is limited.

food. The notions of competition, adaptation, selection, and continuation of species were born independently in the minds of these two scientists. By a lucky coincidence of history, they came together in the mid-19th century when the Darwin and Wallace theory of evolution was first outlined in 1858. It was a British clergyman, the Reverend Thomas Malthus (1766–1834), who in 1798 first spelled out the problems associated with exponential and ultimately unsustainable growth resulting from high success rates in the reproduction of any organism, and these problems are still very much with us.

We have gradually come to realize through our growing understanding of Earth's history, its environments and inhabitants, and the processes that have sustained them that we live on a relatively small planet, the resources of which are not inexhaustible. The Earth's geological history shows that the progress of life through time has not been smooth, but rather filled with vicissitudes, booms, and busts, which at times have drastically cut life back and radically altered the face of the Earth. The driving forces for these changes lie both within and without the Earth and are mostly beyond our control. The best we can do is to learn as much as we can about them in order to better predict their recurrence, take avoiding action where possible, or, if avoidance is not possible, work out how best to cope with the impacts and "mop up" afterwards.

Coral reef
Tropical coral reefs have been the marine equivalents of the terrestrial rain forests for hundreds of millions of years. They support a remarkable diversity of life from simple algae and sponges to vertebrates, such as fish and turtles, many of whose remains can be fossilized, especially those with rock forming skeletons such as corals.

Rainforest

For over 350 million years, from late Devonian times, land plants have formed dense and diverse strands of forest, especially in humid tropical latitudes. Competition amongst the plants for light leads to different growth forms which provide food and shelter for a great variety of animals from microbes to insects and vertebrates.

The problematic record of "deep" history

It took a long time for scientists finally to accept what fossils are: the remains of once living organisms. There were very real problems of interpretation at first. So many fossils are preserved in strange ways with fossilization tending to obscure or remove all traces of original organic materials. Often, fossils retain only a physical resemblance to organic life, being preserved by purely inorganic and often crystalline materials. Even when fossils were recognized as being similar to living creatures, the presence of fossils in rock strata far inland and at the top of high mountains required an explanation.

To begin with, the only possible explanation seemed to be the Flood as described in the Old Testament. Not until the beginning of the 19th century was there any real understanding of the processes of petrification whereby life could be turned to stone. It then took another 100 years and more before the scientific revolution provided satisfactory understanding of the processes which transform seabed sediments and their organic remains into fossiliferous strata which have been folded, faulted, and uplifted to form mountains. Furthermore, it is only in the past decade or so that the unifying plate tectonic theory has emerged to provide a general explanation of Earth processes. For the first time, we can explain how fossils such as those found in coal seams originating in tropical rainforest trees or coral reefs from tropical latitudes come to be incorporated into high-latitude mountains thousands of miles away from where they originated.

At the beginning of the 19th century, scientists were still struggling to come to terms with the collapse of the old paradigms for the history of life and its origins. It is worth briefly recalling just how far we have travelled scientifically over this short time. For nearly two millennia, the world view that prevailed within the scientific community of Europe and the Americas was that of the Judeo-Christian tradition based on a belief in the historical truthfulness of biblical texts with their story of Creation and the Flood. According to this view, life was purposefully designed and created by a benevolent God for the express purpose of supporting the existence of His special creation, humanity on Earth.

Many of the scientists investigating the natural world were firm believers in this tradition, and many were ordained priests who justified their scientific work on the grounds that it would reveal the wondrous details of the Creator's work. But problems soon emerged. For instance, how could a benevolent God allow any of His creations to die out as the fossil record seemed to indicate? Nevertheless, to begin with, the fossil evidence did seem to support the biblical Flood story. By mid-19th century, however, the Ice Age theory had replaced that of the Flood interpretation. Even so, it was not until the end of the 19th century that human antiquity was generally accepted as scientific fact. Some people still do not accept it today.

Scientific investigation of the history of life on Earth, as recorded by fossils, has revealed that life has been abundant for a very long time – hundreds of millions of years, in fact. The further back in geological time we go, however, the less diverse it must have been. We now know that the environments of Earth were not colonized all at once. Life probably evolved in seawater and could not move into freshwater, onto land, then into the air until it was equipped to do so. As new evolutionary adaptations arose, life invaded new ecological territories or "spaces". Increasing populations and separation of populations promoted speciation, or the evolution of different species, with large jumps in overall diversity.

Unfortunately, it is very difficult to recover complete information about these increases in diversity because of the nature of the fossil record. Fossils provide a view of the past that is heavily biased towards particular environments, mainly shallow-water marine ones. Fortunately, there are occasional "windows" where exceptional preservation gives privileged insights into the deep past. Fossil amber, the mineral replacement of soft animal and plant tissues, dehydration, and deep freezing – all have played a role in generating some of these windows.

So far, the known fossil record only encompasses some 200,000 or so ancient species for the whole of "fossiliferous" time. This represents an incredibly small sample of the totality of life. Suppose the average diversity of life has been 10 million species over the past 500 million years and that, with evolutionary turnover, species duration is 10 million years. This would mean that over 500 million years there have been 50 turnovers of 10 million species, producing a grand total of some 500 million fossil species. At present, we only know of some 200,000 fossil species, which is far fewer than 1 per cent and, in fact, only 0.04 per cent of that grand total. Even though palaeontologists still have a large amount of work to do, we will never be able to recover anything like the original total.

Global growth potential
Global biosphere data shows the distribution and abundance of oceanic phytoplankton at the base of the marine food chain (red and yellow representing high concentrations). The terrestrial measurements show potential for production of vegetation with green showing high growth areas and the buff colour representing low growth potential.

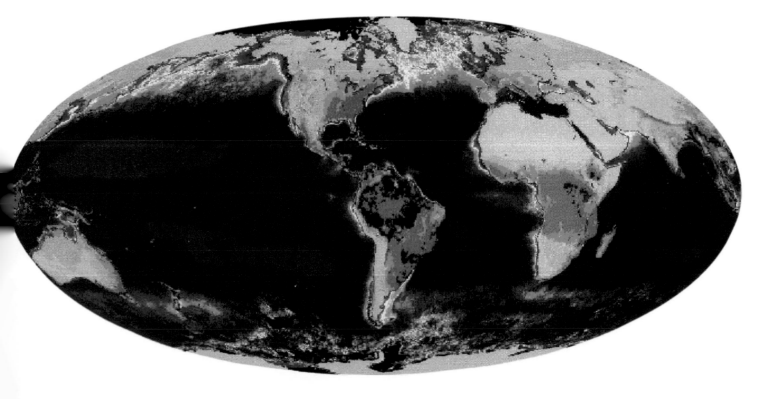

Life's relationship with the environment

Leopard

Natural selection has led to the survival of a remarkable diversity of body forms and lifestyles which are often repeated within different groups of animals and plants. Fleet-footed plant eaters, such as this small gazelle, preyed upon by fast running predators, such as this leopard, have coevolved over hundreds of millions of years.

All organisms are to some extent controlled by their surroundings. Most life can only exist within fairly narrow ranges of temperature, wetness/dryness, supplies of air and preferred food. As a result, life tends to be tied to particular environments. Polar bears and arctic foxes are well adapted to surviving in extreme cold as long as they have enough high-protein food. The African elephant can survive hot, semiarid desert conditions provided it can reach a water supply every few days and find enough plant material to eat in the meantime. They could not, however, swap places and survive for long. Yet, these three animals are all warm-blooded mammals and share a common ancestor who lived some 70 or 80 million years ago. Since that common ancestor roamed the Earth, these mammals have evolved and adapted to survive in very different climates.

If we only had the skeletal fossil remains of these animals to provide us with information, we could not determine the conditions under which they lived. As we know, bears, foxes, and elephants are all big groups of animals that include many different kinds (both at the genus and species level) which are separately adapted to different living conditions without necessarily leaving any trace of these differences in their skeletons. Still, all is not lost. Fossils are found in sedimentary rocks, and most sedimentary strata retain characteristics determined by the environment in which they developed and were deposited. By analysing the sedimentary context and accompanying fossils, especially plant remains, scientists can often deduce the kind of environment in which the animals and plants of the past lived.

The geographical distribution of different life forms varies enormously. Many parasitic organisms depend on just one or two host species for survival, so their geographical range is intimately tied to that of the host. Many flowering plants depend upon certain insects for pollination, and many animals depend on just one or two plant species for food. These are all examples of specialists, the distribution patterns of which are linked. By comparison, generalists such as humans, foxes, and

bears can cope with a large range of food types and consequently can respond quite well to sudden changes in food supply. But humans have not always been so adaptable. It is likely that several human-related species, such as *Paranthropus boisei,* became extinct because they were too specialized in their feeding habits.

The close interrelationship between organisms and their environments and constraining physical and biological parameters raises an interesting problem. How is it possible for life forms ever to leave the "home" environment, the one to which they have adapted, for a new and different one? There is good fossil and biological evidence to show that this has happened time and time again. For instance, several different groups of organisms have at times left the environment of the sea or freshwater for dry land, while others have literally taken flight from land. Life cannot simply wish such changes to happen in order to exploit new vacant plots or spaces with very different physical and chemical conditions. Getting about, breathing, and feeding in a supportive liquid medium such as water is very different from trying to do the same things in a light, dry gas such as air. Adaptation must begin even before such new niches can be exploited. How do such preadaptations come about?

As we are learning to our cost, the environments of the Earth are not as stable as was once thought. The Earth's surface is dynamic and constantly changing, even though the rates of change can be very slow. Changes can also take place at an uncomfortably rapid pace, as we are beginning to realize. Climate, in particular, is a major controlling factor on life. Most plants require certain levels of rainfall, temperature, etc. Consequently, patterns of vegetation distribution tend to be zoned according to climate patterns. Animal life ultimately depends on plants for food, and, if plant distribution patterns are disrupted by changing climate, there can be a cascading collapse through the food chain from the plants to the plant-eating animals and finally to the carnivorous ones.

Investigation of the geological record shows that the Earth's environments have changed drastically in the past and therefore so, too, must have past climates. In northeastern North America and northwestern Europe, the ancient rock succession around 380 million years ago (Ma) shows a sequence from semiarid desert conditions in Devonian times, through shallow seas full of coral reefs in Carboniferous times (c. 340 Ma), followed by rainforests, and back to hot, dry deserts in Permian times (c. 280 Ma). This entire rock sequence is now found in high latitudes that often experience winter snowfalls. When first discovered, these very contrasting data were seen as strong evidence that climates had changes drastically in the past. It is much more complicated than that, as it turns out, because the landmasses have not always occupied the geographical position that they now do.

Nevertheless, climate change certainly has happened – and in the not too distant past – with a series of climate oscillations which we now know as the Quaternary ice ages. These changes had a drastic impact on environments and the life which they support.

Polar bear
The extremes of polar climates have only really been successfully conquered by plants and animals (such as warm blooded mammals and birds) which can adapt to the sub-zero temperatures and limited food resources. Here, the well camouflaged arctic fox and polar bear are both predators that depend upon the superb insulation of their fur.

Processes which change environments over time

Our view of the Earth and the stability of its environments has been radically changed by the discoveries of geology and new techniques of observation. We now have a perspective on the past which spans the entire 4600 million years of Earth's history. Inevitably our view of the recent past is much more detailed than that of remote times. Nevertheless, some major events and episodes in Earth's early history which have impacted upon surface environments and their inhabitants, however primitive, are gradually becoming clearer. We now have good evidence that there were fundamental differences in the composition of the Earth's early atmosphere. Even the remote past was punctuated by ice ages and extra-terrestrial impact events which must have had severe effects on primitive life and its evolution. Oceans were opened and closed, with continental masses growing, breaking apart, and being shuffled about over the Earth's surface.

The study of the biological relationships of organisms to their living environments has shown that most life is adapted to tolerate and survive some change, whether from a twice-daily rise and fall in sea level, a temperature change from day to night, the lunar monthly tidal cycle, or seasonal or annual changes. Even if these familiar rhythms are disrupted or shift in any significant way, organisms have changed either by moving, adapting, and evolving, or by dying out.

With more than 150 years of modern scientific measurement and analysis of living environments behind us, many processes of change are now familiar and reasonably well understood, if not necessarily predictable. Processes which effect major changes on Earth's environments tend to operate on very different time scales, frequencies, and geographical impact from those which operate at local and annual to regional and decadal, or longer, time scales. We are most familiar with the types of changes that are frequent and widespread, such as climate fluctuations operating on an annual or decadal scale, and which can be experienced within an individual lifetime.

Powerful storms and flooding which have enough energy to damage property and even threaten lives are familiar events that occur annually in many parts of the world. Thanks to modern observation and analytical techniques, we have a better comprehension of how, in any one location, the magnitude and frequency of these weather phenomena may change as climate changes.

Similarly, earthquakes and volcanic eruptions are frequent in some regions of the world, especially around the Pacific rim from Patagonia to New Zealand. We build cities in earthquake zones and within the destructive shadow of active volcanoes because of the pressure for living space and fertile land. Again, life tends to accommodate to such processes, but the geological record shows us that repeatedly in the past much larger scale but rare events have happened and will happen again, especially volcanic eruptions. There are vast regions of India, Siberia and northwestern United States which are covered with great thicknesses of lavas known as plateaux or flood basalts that poured out of fissures in the ground and literally flooded surrounding landscapes, destroying life. As no such events have happened within recorded history, it has taken a long time for scientists to realize that they do happen.

Destructive tsunamis (tidal waves) can be generated offshore by earthquakes, submarine volcanic eruptions, and slumps of huge masses of seabed sediments. These

Hurricane Mitch

We now know that rare but exceptionally powerful storms and unusually extensive flooding can happen on decadal (10-year) and centennial (100-year) frequencies. Hurricane Mitch, which occurred in October 1998 and killed some 11,000 people, became known as the storm of the century, and it was certainly the worst to hit the Americas since the 1780 hurricane which probably killed more than 22,000 in the Caribbean region. The effect of such rare events on environments can be more devastating than all the intervening annual events.

events displace huge volumes of water locally and send shock waves across oceans to wreak havoc thousands of miles away. Until recent decades, most scientists (except those in Japan, which had direct experience of such events) had little knowledge of this phenomenon. We now suspect that tsunamis perhaps as much as a kilometre high have been generated by the impact of large, extra-terrestrial bodies in the oceans. Ancient sand layers, some 65 million years old, found around the Gulf of Mexico are thought to have been deposited by giant tsunamis generated by the Chicxulub event which marked the end of the Cretaceous period and the final

demise of the dinosaurs. Again, fortunately, no such large-scale event has happened within recorded history. It is enough to know of the destruction to the Portuguese capital of Lisbon in 1755, and the deaths of 70,000 people, when an earthquake hit the city and also generated a tsunami that compounded the destruction.

Devastating landslides were once thought of as relatively rare events. We now know, however, that they are common in geologically young mountain ranges such as the Himalayas and the Andes, and even in less dramatically mountainous regions such as California. Landslides can be generated by earthquakes or heavy

Sand dunes

Terrestrial landscapes are normally being worn down to sealevel, under the influence of gravity by the processes of weathering and erosion. However, there are also some important environments such as deserts where there is a net accumulation of sediment which stands a considerable chance of being preserved in the future rock record.

Changing climate

Satellite observation has recently shown significant changes in and the break-up of major ice sheets, such as that of the Larsen Ice Shelf, the largest on the east coast of the Antarctic Peninsula, due to global warming. The exact effect of the release of such large quantities of cold ice on the life of the surrounding seas is not yet known.

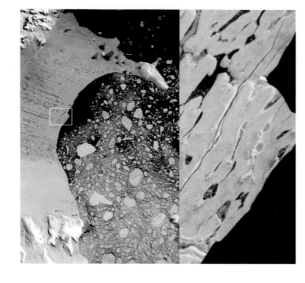

rainfall, and they are common events wherever this combination occurs in seismically active, hilly terrains such as the monsoon-drenched regions of Southeast Asia. These processes have a considerable impact on the environment and its inhabitants. Under the influence of gravity, rock and soil debris avalanche downslope into streams and rivers, which carry the debris away and dump much of the load on their downstream flood plains. The hillside scars expose fresh rock to the slower processes of weathering, soil formation, and invasion by vegetation before the whole cycle happens again. The net effect is constant turnover and transport of Earth materials, part of an overall recycling. But if climate change were to take

place in the form of, for example, cessation of the monsoon, there would not be enough rain to support such profuse vegetation and soil development, and the environment would not be able to support so much life.

Just how dramatic the environmental effects of climate change can be is best seen around the margins of arid regions such as the Sahel in North Africa. The occasional rains and sparse scrub vegetation of semiarid regions are often just about enough to support migrant populations of game and people with herds of domesticated cattle. If those sparse rains fail, however, even the desert-adapted plants will die or be used up by desperate people, with the result that the environment becomes uninhabitable. Without plants, the thin soils deteriorate, and the sediment is more prone to wind erosion and the encroachment of shifting dunes. Interestingly, there are recent reports suggesting that some of the desertification has been reversed.

We humans have notoriously not been good "stewards" of the land. In the past, much of this has been through ignorance of the "knock-on" effects. The overexploitation of the prairie soils of North America's Midwest in the early decades of the 20th century is a good example of this, with increasing mechanization allowing vast acreages to be ploughed for planting grain. Hot, dry summers and soils unprotected by the native grasses led to wind erosion, the development of "dust bowls", and eventually badlands. We have no such

Extinct birds

One of the most fascinating images of the submerged Cosquer cave is this charcoal outline of a bird, first thought to be that of a penguin but now recognized as an accurate image of the extinct penguin-like flightless great auk which was finally hunted to extinction in the mid 19th century. Fossil remains of the great auk have been found around the Mediterranean whilst penguins are inhabitants of the southern hemisphere of the globe.

excuses today because the vulnerability of plant cover, soils, and whole environments is now well understood; however, population pressure continues to have a deleterious effect in many regions. Virgin forest is still being cleared, first to provide timber, then space for agriculture. Not only is there a serious loss of habitats for many endangered species, but also these forests are thought to act as carbon dioxide "sinks" which help to ameliorate the effects of global warming.

Sea-level changes were a constant feature of the distant past and can result from a number of geological processes. The best known is the kind of drastic climate change associated with ice ages. So much ice is produced that significant quantities of water derived from the oceans are locked up as ice on land, and global sea levels drop. Global warming heats oceanic waters sufficiently to cause them to expand, adding to rising sea levels. When the ice eventually melts, the process is reversed, but the slow depression and rebound of the ice-laden land complicate the net effects. For instance, at present Northern Europe is still rising due to rebound after the melting of the Quaternary ice sheets, despite the fact that sea levels are rising due to global warming. Land movements on a local and regional scale can also produce relative movements of land to sea level. Finally, the relative rates of ocean floor spreading can affect global sea levels by changing the volume of ocean basins.

Falling sea levels create land bridges such as that between Asia and North America, which allowed the migration of animals, including humans in the prehistoric past, between the two great continents. Rising sea levels isolate landmasses and their inhabitants, from small islands to continents. The repeated isolation of North America from Asia has had a huge effect on the migration, population isolation, and subsequent evolution of mammals within these continents. Britain and Europe, and Australia and Southeast Asia have similarly been isolated by rising sea levels, which also led to the loss of important productive lands marginal to the sea around the world. Shallow continental-shelf seas flood into low-lying continents and can in turn become highly productive in terms of marine life. The interplay of rising and falling sea levels has been a very important process in the geological past, eg in the development of the coal measure forests of Carboniferous times.

One major process which affects environments in the long term is that of plate tectonic movement, which has opened and closed oceans, moved continents and broken them into pieces, only to reassemble them in new patterns. This process sees margins of colliding plates being crumpled together, thickened by tens of kilometres, and folded and faulted to form major mountain chains such as the Himalayas. The migration of these plates has carried some life along with it, but it has also destroyed an enormous amount of evidence about life in the oceans of the past.

Volcano

Although generally perceived as destructive, volcanoes are often constructive geologically. Their rock products (airborn rock fragments, magma, ash and lavas) often cover and preserve surrounding landscapes and sometimes the remains of animals and plants. When cooled and lithified such volcanic rocks often prove extremely resistant to weathering and erosion.

Earth— a dynamic system in constant flux

From space, the Earth is a blue planet with plenty of water shrouded by a patchy, white cloud cover showing the presence of a humid atmosphere, below which landmasses coloured green, brown, and yellow can be glimpsed. By stark contrast, Earth's satellite, the Moon, is a barren and heavily pockmarked sphere of rock and dust with no signs of life. And yet we now know that the Moon is some 4000 million years old, only 600 million years younger than the Earth. So why do these two bodies look so different? Why is the Earth not just as craterous as the Moon?

Indeed, all planetary bodies are subjected to bombardment by objects of all sizes ranging from a daily rain of dust to large asteroid sized rock masses every few million years. The planetary bodies in the Earth's solar system are also subjected to temperature change because of their orbits, axis of rotation, and rate of spin relative to the Sun. Those with gaseous atmospheres have winds as well, and, if there is any water in the system, there is the possibility of ice and ice caps, steam, clouds, rain, and oceans. The state of the moisture depends, however, on the ambient temperature, so that Europa, one of Jupiter's moons, has a surface temperature of minus 170 degrees and is covered in ice. But it is evident that this ice cover is remodelled in a cycle which starts again every 10 million years or so.

If the Earth were "dead" as the Moon is now, its surface would look like that of the Moon, frozen in time and still pockmarked with craters from ancient impacts and volcanism that produced vast ancient lava "seas". The Moon has no internal heat energy, no atmosphere, and no water. Although the Earth has been subject to events similar to the ones the Moon has, the constant recycling of the Earth's surface has destroyed much of the evidence. Luckily for us, the Earth is still very much "alive" and dynamic, otherwise there would be no life on our planet. There are several reasons for the Earth's continuing vitality. Its distance from the Sun, its orbit, its axis of rotation, and its rate of spin – all help to produce a relatively equitable surface environment. It is neither too hot to burn off all the moisture nor too cold to freeze it completely. There is enough water to generate an atmosphere and oceans, which further help to cushion us from the extremes of solar radiation and their potentially harmful rays (especially ultraviolet light).

Most importantly, our planet has an internal source of heat that drives the great Earth machine with its continual convection heat flow mechanism. This ongoing mechanism creates and destroys oceans and drives the movement of the vast crustal plates into which the surface of the Earth is fragmented. The action and processes of convection currents and plate movements are large scale and long term. While the mechanisms are still far from being fully understood, there is incontrovertible evidence that plates can move over thousands of kilometres over tens of millions of years. This may seem extraordinary, but spreading rates of 10cm (4in) per year are equivalent to a kilometre every 10,000 years and 100km in a million years. Subducted slabs are recycled within the Earth over even longer periods – probably hundreds of millions of years.

Overall, the effect has been to give the Earth a complex and dynamic history which geologists are still trying to unravel. The most recent geological past is the most familiar; the further back in time we proceed, the less clear the story becomes. The past 200 years of modern scientific investigation have given us a fairly good idea of the outlines of Earth history back over the past 500 million years or so. Beyond that, the much greater stretch of Precambrian time is still poorly known, but is the subject of active investigation.

A general look at the main topographical features of the Earth's surface provides clues to the patterns and mechanisms of our planet's internal heat machine. In comparison to our human scale, variations in the Earth's surface topography seem to be extreme. Mountains rise so high above sea level that temperatures are permanently below freezing and oxygen is in short supply at their peaks. (Everest, for instance, is nearly

9km (6 miles), or 8846m (29,022ft), high.) They are so inhospitable to life that humans can visit them only briefly; otherwise they are normally devoid of life. These high mountains typically occur in elongate ranges extending over hundreds or thousands of kilometres. Geological exploration of them has shown that each mountain belt has a particular history related to very large-scale processes. Their rocks are folded and faulted by massive compression, thickening the crust to such an extent that their "roots" have partially melted, producing volcanoes and large intrusive rock bodies.

The global distribution of major earthquake zones with their related faults, volcanoes, and eruptive products all provide clues as to underlying dynamic processes within the Earth. Geologists study these phenomena today and relate their findings to evidence for the same processes which occurred in the past. As a result, the "deep" geological history of the Earth has been mapped out and is now quite well established for the past 500 or so million years, and this history is gradually being extended deeper and deeper into the more ancient past.

Hurricane force

Cyclonic storms, such as Hurricane Andrew seen here in the Gulf of Mexico in August 1992, are normal seasonal events in many tropical regions. In recent decades it has become evident that every hundred years or so superstorms of even greater magnitude can occur, which have the potential to leave lasting marks on the rock record, especially in continental-shelf sea deposits.

Landslide

We humans tend to be rather shortsighted, and persistently underestimate the power of the natural processes of weathering and erosion. Hillslopes and coastal cliffs might seem perfectly strong and longlasting when dry. However, in the long term they invariably crumble away under the combined effects of heavy rains, slumping, sliding and avalanching of rock and sediment.

The greatest revolution in the understanding of the Earth's dynamism has resulted from exploration of the oceans over the past few decades. While details of the Earth's land topography have been known for many years now and carefully mapped by surveyors, mapping of the bottom topography of the oceans (some 70 per cent of the Earth's surface) has been a much more recent achievement. It has depended upon sophisticated ship-borne techniques and more recently the same type of remote satellite sensing that has been applied to the mapping of the land, its topography, soils, vegetation, and structural features.

Major topographical features have been discovered which tell us a great deal about the inner workings of the Earth. Some of the most dramatic topographical features are the ocean deeps such as the Marianas Trench, which extends 11,033m (36,198ft) below sea level. There has to be some deep-seated "reason" for the ocean floor suddenly to descend to such depths, some force has to be pushing it down, a force that can now be explained by plate tectonics.

The difference between these depths and the highest mountains gives an overall vertical range of nearly 20km (12 miles). Yet most of the land surface has no great elevation. Indeed, the average height of the land is a mere 840m (2757ft) above sea level, while the average depth of the oceans and seas is 3.8km (12,460ft) below sea level. Not only are the oceans significantly deeper than the land is high, relative to sea level, but also the area of the Earth surface taken up by the oceans is much greater than that of the land (70 per cent). The oceans dominate the Earth's surface and have done so for billions of years. Moreover, most of the history of life has been in the oceans. Life only set "foot" on land for the first time around 460 million years ago.

This variance in average height of the land and the depth of the oceans is caused by an underlying fundamental difference in the rock materials from which they are made and has a far-reaching implication in major geological processes. Basically, the oceans are floored by igneous rocks that are denser than the rocks which make up the continents. The ocean floor is not just marked by deep ocean trenches. It also features the longest mountain chains on Earth. One of these oceanic

mountain chains runs southwards in mid-ocean from Iceland down the length of the whole Atlantic, before diverging with one branch extending eastwards into the Indian Ocean and the other extending westwards to the southern tip of South America. Compared with the more familiar mountain ranges on land, these submarine ranges have some interesting and important differences. Not only are the ranges much longer, but they are symmetrical in cross-section as well, and it is no accident that they are known as mid-ocean ridges because they are also spreading ridges which mark the site of the ocean growth. Why are there mid-ocean mountain ranges, and why are they different from mountain ranges on land? Again, plate tectonic theory is providing an explanation.

Other major features of the ocean floor include chains of volcanic ocean islands such as the Hawaii–Emperor chain which extends thousands of kilometres from mid-Pacific northwest to Kamchatka. The islands of Hawaii are one of the biggest and most active volcanic complexes on Earth, in that they rise some 5000m (16,400ft) from the ocean floor, then a further 4000m (13,000ft) above sea level. Northwest from Hawaii, the islands of the same chain become less active, less high and geologically older. Eventually they do not even rise above sea level, but continue as a line of sea mounts that sank beneath the waves as they cooled. Their existence and nature are signs of an ocean crust plate moving over an active "hot spot" deep within the Earth. As long as the Earth is a dynamic system in constant flux, such processes and events will continue. They may be difficult to live with, but without them the Earth would "die".

Isostasy

Under the principle of isostasy continental masses "ride" high on the denser rock of the ocean floor and the semiplastic rocks of the Earth's interior (the mantle) below, similar to an iceberg floating in the sea. As icebergs melt, or continents are worn down, they continue to rise, maintaining the same position relative to sealevel as the material they displaced returns beneath them. Icebergs float with about 10 per cent of their mass above water. Similarly, continents, which have an average thickness of around 35km (21¾ miles), show only a small proportion above sealevel.

Environments today as potential sites of fossilization

If you have a particular desire for immortality and fossilization of your mortal remains for hundreds of millions of years, then you have something of a problem. Burial on land can work. After all, the fossil record of dinosaurs largely depends upon it, but the vast majority of fossils are of marine organisms. To be preserved as fossils, organic remains generally need some preservable hard parts and have to be buried deep within the Earth. Unfortunately, landscapes do not normally make good long-term burial sites because they are continually eroded, with their rock debris eventually being dumped at sea by rivers or wind.

The vast offshore sites of sediment accumulation which form the shallow water shelves around the continents are the main burial grounds for potential fossils. These layers of sediment build up into piles many hundreds or even thousands of metres thick and over time become compacted into sedimentary strata. Burial in the deep oceans is as problematic as burial on land, however, because most ocean floor is eventually subducted by the processes of plate tectonics. But the thick piles of sea-shelf sediment fringing the continents do not subduct so easily. Rather, plate collision deforms and uplifts these vast aprons of sediment to form new mountain ranges for future geologists to hammer away at, looking for fossil evidence to find out their age and environments of deposition.

So, for any remnants of ancient environments to be preserved in the rock record, various conditions have to be met and a succession of events has to take place in the right order. In general, the right combination and sequence are not found as often as one might expect. If we pause to consider the dynamic nature of the most common of Earth's surface environments, however, it soon becomes clear why this is so. Landscapes are for the most part above sea level. All the processes of weathering and erosion impact upon them. Under the influence of gravity, any loosened surface material, whether carried away by water or wind, essentially moves down slope and is eventually carried to sea level.

This overall trend may take a very long time – many millions of years in some instances – but it is fairly inexorable. In the long term, landscapes are worn down by this process of peneplanation, as it is called, to a base level close to sea level. Even the highest mountains can be worn down and removed, with their debris dumped in the sea, over a very long period of time. The fact that we can see igneous rock such as granites, which were originally intruded at depths of tens of kilometres and are now exposed at the surface, tells us that tens of kilometres of surface rock material can, with sufficient time, be stripped away. However, there are important geological exceptions to the general trend of removing sediment from land and dumping it at sea. The most important is the formation of sediment traps within continents, either within topographic or structural basins.

Large topographic basins of sediment can form where high rates of sediment accumulation combine with subsidence of the crust over a long period of time. Rapid sediment deposition tends to occur on the flanks of rising mountain chains where the crust is stretched and thinned, and tends to sag. High rates of erosion and deposition lead to the development of large alluvial fans, river flood plains, and, sometimes, lakes or inland seas as well. The accumulating weight of sediment further depresses the crust and allows thousands of metres of deposits and their contained fossils to build up. Such depositional environments develop on the flanks of young mountain belts such as the Alps and Himalayas. Rapid uplift of the mountains promotes increased rates of erosion (with land surfaces being reduced by over 6mm (¼in) per thousand years), transport and deposition of sediment on the surrounding lowlands. For instance, the Ganges–Brahmaputra region south of the Himalayas is a site of vast sediment accumulation with over 1500 tonnes per square kilometre deposited each year. Such thick sequences of continental strata have a good chance of being preserved in the rock record, as do those that accumulate in rift valleys.

From the human point of view, the role that rift valleys have played in preserving strata and fossils is very important. This is due to one rift valley system in particular, the Great Rift Valley of East Africa, a remarkable feature so large that it can be seen from space. The valley stretches some 8000km (5000 miles)

from Mozambique northwards through Kenya to Ethiopia, the Red Sea, and right up into the Jordan Valley.

Rift valleys originate where the Earth's crust is being stretched. As rocks are mostly brittle, rather than elastic, they tend to break along fault lines. Between parallel faults, the intervening rocks sag or collapse like the keystone of an arch that fails through lack of side supports. The downfaulted central valley then becomes a site where sediment can accumulate over a long period of time and stand a good chance of being preserved even when the surrounding landscapes are worn down and removed by erosion and weathering. In the example of the Great Rift Valley, evolutionary, geological, and geographical accident or coincidence have luckily combined to preserve a record of much of what has happened over the past 15 million years or so. Both sedimentary environments and a remarkably good record of the life of that period can be found in these rocks. Through a fortunate piece of timing, this just happens to coincide with a critical phase in the evolution of early humans and our extinct relatives. If ever there were an Eden, this is where it was. Charles Darwin was right: a critical part of our history and evolution is to be found in Africa. Indeed, genetic studies have recently shown that all living humans are essentially African in ancestry.

Pompeii
The eruption of the southern Italian volcano of Vesuvius in AD 79 produced a number of catastrophic flows of hot gas and ash, known as pyroclastic flows, which rolled down the mountain overcoming the towns of Herculaneum and Pompeii, suffocating many of the citizens. Such catastrophic events which preserve an instant in time are known today as "Pompeii events".

Underwater plane wreck (far left)
Examination of the successive colonization of submerged and accurately dated historic wrecks gives useful measures of the rates at which marine organisms grow and at which remains are buried within seabed sediments.

Our view of the geological past

Grand Canyon

America's Grand Canyon is one of the few places on Earth where it is possible to see a sequence of strata stretching back over a 1000 million years from the Cenozoic era right back into the Precambrian eon.

We now know that the Earth has a lengthy history extending back around 4600 million years. Life on Earth also turns out to have had a remarkably lengthy history, with fossil evidence appearing in the rock record perhaps around 3800 million years ago and certainly by 3500 million years ago. Strangely, it also appears that it took a very long time for life to "get going" in its evolutionary cycle. Much of its early history was microbial, and it took a long time for multicellular life to evolve, perhaps as much as 1500 million years. Increase in size and biological complexity to a level where fossils became readily visible to the unaided eye took even longer. Not until late Precambrian times around 600 million years ago, some 3000 million years after life actually began, do we begin to find what most of us would think of as fossils. Even then, these creatures were all still bound to the seas, and at this stage were entirely soft-bodied. The evolution of hard shells and skeletons had not yet

begun. So life seems to have had this extraordinarily long "slow-burning fuse" within Precambrian times.

Part of the reason for the slowness of early evolution may have been the changes that took place in the early development of ocean and atmospheric chemistry. We know that there was something of a "chicken-and-egg" problem in the oxygenation of the early atmosphere. Degassing from early volcanic activity provided the first atmosphere, and it was dominated by carbon dioxide (about 80 per cent); initially, there was no free oxygen. Gradually, moisture in the atmosphere was cracked into its components, oxygen and hydrogen, resulting in the slow build-up of oxygen and the escape of the lighter hydrogen into space. Early sedimentary strata rich in chert, which forms mainly in low-oxygen environments, bear witness to this state of affairs. The slow oxidation of the atmosphere is also recorded in the accumulation of iron-rich sediments called banded iron formations which today provide most of the world's iron ore. It appears that it was not until around 1800 million years ago that oxygen appeared in any significant quantities in the atmosphere and allowed the evolution of organisms that "breathed" oxygen.

The subsequent history of life over the past 500 million years and more has been well known, at least in its broad outlines, since the early 1900s. The biologist Peter Medawar (1915–1987) thought that this outline was so well known that any further investigation of fossil life was merely a matter of "dotting the i's and crossing the t's". As we shall see, he was wrong, unless you consider the discovery of feathered dinosaurs as merely a case of "dotting an i".

Two hundred and fifty years of scientific exploration have shown us that we have a geological rock record accessible at the surface of the Earth. Surprisingly, this record extends back almost as far as the origin of our planet. It took most of the 19th century for geologists to carve up geological history into a succession of identifiable periods of time such as the Carboniferous and Cambrian. The efforts of explorers and surveyors who discovered new lands, lakes, and mountains are often celebrated and well known. Yet who knows of the Reverend Dr W. D. Conybeare and his colleague W. Phillips who named the Carboniferous period in 1822, or of the Reverend Professor Adam Sedgwick who named the Cambrian period in 1835?

This mapping of the temporal dimension of Earth history through its rocks was a remarkable achievement, one that few people fully appreciate. Long before geologists even believed in evolution, they realized that the fossil content of successive strata changed and that fossils could be used to distinguish strata. The complication is that many strata are folded and faulted by Earth movements and that there are large areas of the land surface where no rocks are exposed. Nowhere on Earth do we find anything like a complete sequence of rock strata representing all the different phases of Earth history. Only by a long and complex procedure of mapping and dating rock strata, and comparing their sediments and fossil content from different sequences around the world can a composite log record of geological history be compiled. In any one location, the local sequence of strata will be full of gaps when no sediments were recruited to the record.

Darwin was right when he complained of the incompleteness of the geological record. In fact, the gaps in the record are often telling us about important events other than deposition, notably processes that have led to uplift and erosion, often associated with mountain building and related to plate tectonic movements. The investigation of the huge range of Precambrian time is one of the last great unknowns of geology. Much has been discovered – its life and the existence of major changes in environments over these 4000 million years, including astonishing ice ages – but so much more remains to be discovered.

Insect in amber
The preservation of past life as fossils varies enormously in its quality and completeness, from the occasional broken bone, shells, or footprints, to the almost perfect preservation of organisms trapped in amber resin. However, despite appearances and past claims, amber does not preserve ancient biomolecules such as DNA.

2
The ice age

The gradual realization that an Ice Age, rather than the biblical flood, scattered sediment and the fossils of large mammals across northern Europe and North America opened a new window on the prehistoric past. One kind of catastrophic event was replaced by another just as extraordinary and with far reaching consequences for humankind. There was mounting evidence for glacial activity as far south as New York, Birmingham and Berlin. A Swiss geologist, Louis Agassiz, was even arguing that a global glaciation might have reached as far south as the Amazon. How and why had it happened and what effect had there been on the life of the times? Were humans exposed to such drastic climate changes? The mid 19th century was a period of scientific turmoil with a constant deluge of new discoveries about the natural world and its prehistory. There were intense debates over the interpretation of the finds. Could they be fitted into a Lyellian vision of gradual change over enormously long periods of geological time, or was the older idea of the role of catastrophism making a comeback?

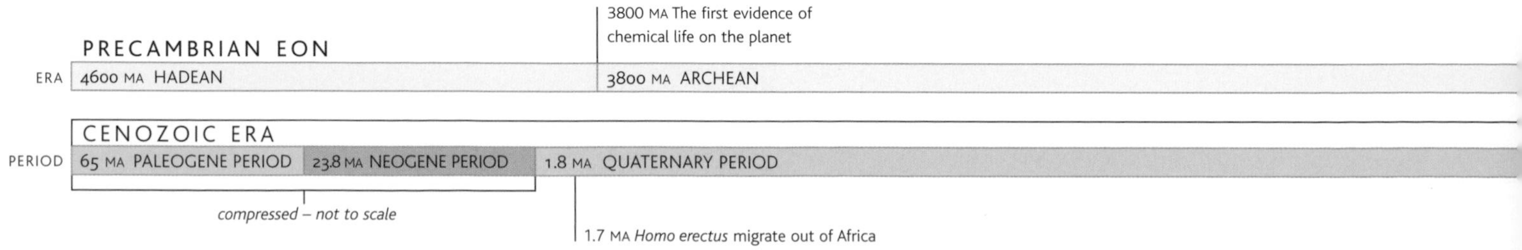

PRECAMBRIAN EON

ERA | 4600 MA HADEAN

3800 MA The first evidence of
chemical life on the planet

3800 MA ARCHEAN

CENOZOIC ERA

PERIOD | 65 MA PALEOGENE PERIOD | 23.8 MA NEOGENE PERIOD | 1.8 MA QUATERNARY PERIOD

compressed – not to scale

1.7 MA *Homo erectus* migrate out of Africa

The whole question of Ice Ages and what causes them has taken on considerable importance over the last decade or so. Climate change has become increasingly important as the full implications of what even a few degrees of temperature change over several decades will mean to life on Earth have become apparent, and widely publicized.

The acceptance that our human activity does and will continue to impact upon climate change has spurred investigation of the mechanisms involved.

Furthermore, there has been the sobering discovery that very rapid climate change has happened during the recent Ice Age and may well happen again in the not too distant future. Again the questions of how and why have these changes come about?

There is a considerable urgency today to understand the very complex global climate system with its interlinking of ocean and atmospheric circulation. In the 21st century, we are already witnessing the effects of gobal warming.

Greenland ice sheet

Ice is a remarkable material which can act as a solid but can also flow in a semi-plastic state. Here this ice sheet flows through gaps in mountains to spread onto lower ground. Such behaviour explains the power of glaciers and ice sheets to grow and cover huge areas in relatively short periods of time.

1200 MA The first multi-celled organisms date from the middle of the Proterozoic period

610 MA The first large marine animals appear

PHANEROZOIC EON

| 2500 MA PROTEROZOIC | 545 MA PALEOZOIC | 248 MA MESOZOIC | 65 MA CENOZOIC | TODAY |

150 KA Penultimate glacial maximum
120 KA Last interglacial
100 KA Modern *Homo sapiens* migrate out of Africa
80 KA First bone tools. Graphic symbols in Africa
60 KA *Homo sapiens* arrive in Australia
c.20 KA *Homo sapiens* arrive in North America

30 KA Extinction of the Neanderthals
18 KA Last glacial maximum; lowered sea levels

12 KA Last retreat of ice caps and glaciers
10 KA Human-induced extinction of big game
7 KA The beginning of agriculture
4.8 KA Extinction of last mammoths
Little Ice Age 1500–1800 years ago

The present as a warm interglacial

A lucky accident of geological and climatic history has meant that, over the past 8500 years or so (most of the Holocene epoch), our ancestors have enjoyed a relatively stable and warm climate. New data shows however, that the climate was not always as stable as previously thought. There were significant fluctuations that impacted upon humans. Investigation of the most recent geological deposits on land and at sea and the record of ice cores shows that annual average global temperature has mostly been above 15°C (59°F) during the Holocene epoch. This compares with an average of around 11°C (52°F) for the preceding 5000 years. An interesting question which we shall return to is whether the present warm interglacial marks the end of the Quaternary ice ages or whether, despite global warming, it is just another warm phase before glacial conditions set in again.

Ice caps are present at Earth's poles today, but they are relatively small. It was thought that until the very recent phase of melting and retreat which has taken place since the early 1990s, the ice caps and glaciers

had not fluctuated in size since the last major retreat, which began about 12,000 years ago. This time span coincides with the growth of historical records. These records, along with data from ocean sediments and ice cores, reveal important and rapid climate swings which included temporary regrowth of the ice fronts.

Ten thousand or so years ago, modern humans were still hunters who lived in small bands spread around the world. The global human population was probably still only a few million and did not expand greatly until humans changed their way of life to the more sedentary practice of agriculture around 7000 years ago. The postglacial melting of the great polar ice sheets opened new corridors for migration. Animals and humans moved from Asia into the Americas and from the European continent into the British Isles. These land routes were available as long as sea levels remained relatively low; however, as the meltwaters from the ice caps found their way back to the oceans, so sea levels began to rise inexorably. For some populations, there was no return from their offshore islands.

It is no coincidence that many cultures, apart from those of Africa, contain flood legends that record catastrophic natural floods in prehistoric times. The melting of the ice caps released enormous volumes of water, especially into the continental interiors of North America and Asia. Much of this floodwater was unable to drain away, but instead formed vast inland lakes and seas such as the Great Lakes in North America. Landscapes were radically changed, as was their vegetation and wildlife. Remnants of these postglacial environments can still be seen today in the few remaining wildernesses of North America and northern Eurasia. Extensive evergreen coniferous forests, with fragments of cold steppe grasslands, still support large herds of caribou, as they have done since the end of the last Ice Age, about 12,000 years ago.

Some scientists have argued that vegetation and climate change were most likely responsible for the way in which the large populations of medium to large plant-

Hippo skeleton
The discovery of the well-preserved skeleton of a hippopotamus within Ice Age deposits near Cambridge, England showed that at times the climate oscillated widely from cold glacial to warm interglacial conditions. Sea levels also fell sufficiently to reveal land bridges and allow such animals to migrate into the British Isles from continental Europe.

eating mammals crashed all over the world. Certainly many of these herbivores, such as the mammoth, giant deer, woolly rhino, bison, and wild horse, were grazers dependent upon extensive grasslands for their food supplies. Increasing humidity and rainfall soon meant that these grasslands were replaced by forest and woodlands, environments which were more suitable for browsing herbivores, especially deer. But it has to be more than mere happenstance that crashes in populations of these so-called megaherbivores matched the arrival and establishment of growing populations of modern human hunters. And this happened all over the world except in Africa and parts of Asia.

The present view is that it was primarily the arrival of human hunters that caused the megaherbivore populations to crash. With diminishing game available for food, modern humans were increasingly forced to rely more on gathering plant foods and domesticating some wild animals and plants. While the beginnings of agriculture allowed bigger groups of people to live together and the development of settlements, it also required the land to be cleared of forest and woodland. The success of this first agricultural revolution was probably instrumental in promoting population growth

in the more fertile parts of the world. The process still continues today, with little of the postglacial forests, woodlands, and their animal populations remaining. While the rainforests of the tropics may still seem vast, they are only a shadow of their former extent.

Records of recent climate change

It had been thought that the effects of climate amelioration were mostly felt in high latitudes, but it is now known that the tropics experienced major changes as well. Most important was the greening of the Sahara with woodland, lakes, rivers, and abundant wildlife from around 10,000 to 5500 years ago, produced by warmer and wetter conditions. Monsoons circulated rain over the northern tropics and equatorial East Africa. Cooler and drier conditions and the advance of the tropical mountain glaciers followed. These changes are thought to be due to changes in the Earth's orbit around the Sun, which in turn impacted upon atmospheric and oceanic circulation patterns.

Reliable historical climate data is only available for the past 150 years or so, and then only for a few locations, mostly in western Europe. Nevertheless, these data show some interesting and significant trends. For

Cup and ring

These distinctive cup and ring marks were ground into a rock surface by Neolithic people around 5000 years ago in the Orkney Islands of northern Scotland, for reasons as yet unknown to us. They cut across older grooves ground by glacial ice during the Quaternary Ice Age.

the 50 years between 1860 and 1920, the global annual average surface temperature fluctuated by less than 0.5°C (32.9°F). When the average of these 50 years – 15°C (59°F) – is taken as a baseline and compared with more recent years, there is a clear upwards trend which has accelerated since 1980 and now has increased above the baseline by 0.8°C (33.4°F). The global retreat of glaciers verifies this trend. In fact, some may disappear altogether, such as those of Mount Kilimanjaro, which are predicted to melt away by 2015. This trend provides convincing evidence for the existence of global warming at the present.

Records also show that there was a Medieval Warm period from about 1050–1375. In South America, it produced a 300-year drought which may well have destroyed the Tiwanaku civilization around Lake Titicaca, Peru. A well-documented cold phase, known as the Little Ice Age, followed, beginning in the late Middle Ages and lasting for some 300 years until the 19th century. This produced serious crop failures in densely populated regions such as Europe, leading to famine, disease, and civil unrest, culminating in the French Revolution.

Landscape paintings of northern Europe from the 16th and 17th centuries often show frozen rivers and canals with jolly peasants enjoying themselves skating and even holding fairs on the ice. Very rarely are winters cold enough to produce such freezing conditions these

days. These paintings provide striking visual evidence that, over this period, the climate was significantly colder. Such paintings also tend to put a positive gloss on an otherwise socially disastrous period. In upland and marginal areas such as the Scandinavian highlands, Greenland, and Iceland, land was abandoned and populations dwindled. Yet annual average temperatures only declined by less than 2°C (35.6°F).

Wine cultivation, which had been introduced to Britain by the Romans, was abandoned in this region in the middle of the 15th century because of the deteriorating climate; however, it has been renewed since the 1960s. In Europe, data about the wine harvest since the end of the 16th century are remarkably good. Winemaking communities all over France and in some other countries have regularly recorded how many days after 1 September the grapes were ready for harvesting. The warmer and sunnier the growth period, the nearer to 1 September the grapes were ready. This provides a rough proxy measure of climate and shows constant fluctuation with a range of some 20 days every few years. There are also some longer term trends lasting a decade or so.

To gain more accurate details of climate change, however, other biologically based proxy measures can be used. Basically, any organisms which are abundant and sensitive to climate change and can be easily fossilized have the potential for providing such measures. These include trees and their growth rings, which give evidence for climate change over the past 11,000 years; however, well-preserved timber of this kind is rare. Much more common and useful is fossil plant pollen that can be specifically identified and linked to a local flora of a particular time frame (dated by radiocarbon methods which are effective for the past 50,000 years).

The temperature and climate tolerances of living plant species are well known. From these, past climates can be reconstructed as far back as those species have been in existence, which amounts to a few million years for many species. Some animal fossils can be used in the same way, especially certain insects and particularly beetles whose tough carapaces are often preserved in ancient peat and lake deposits of the recent ice ages. More important, however, are certain deep sea-dwelling microorganisms, known as foraminiferans, or forams for short. Analysis of their fossil remains recovered from deep sea sediments has allowed scientists to measure relative changes in the volume of the oceans, which in turn are related to climate and temperature changes. These detailed analyses have been supplemented by data derived from ice cores drilled through the polar ice caps

that have built up over the past 300,000 years or more.

Until about 6000 years ago, human activity impacted mainly upon the biological world. Until this time, tools were derived predominantly from natural organic materials such as wood and bone, and inorganic materials such as flint or obsidian rock materials, from which blades and axe heads were made. The recent discovery of a well-crafted copper axe hafted with yew and dated at around 5000 years old shows, however, that metal extraction from mineral ores must have begun by this time. The axe was found with a frozen body (nicknamed Otzi, the Iceman) in a glacier on the Austrian–Italian border in 1991. Further, the 4000-year-old level within the Greenland ice cores shows a small but distinct rise in dust particles derived from the smelting of copper and other base materials. This demonstrates that metalworking had risen to such an extent that it was already beginning to cause atmospheric pollution. We have escalated this type of pollution to such an extent that there is little doubt that it is contributing to global warming. The stability of the Holocene climate inherited by our ancestors is probably now at an end.

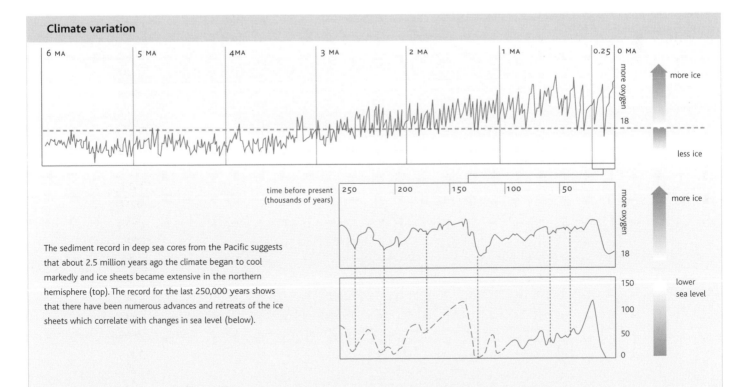

Climate variation

The sediment record in deep sea cores from the Pacific suggests that about 2.5 million years ago the climate began to cool markedly and ice sheets became extensive in the northern hemisphere (top). The record for the last 250,000 years shows that there have been numerous advances and retreats of the ice sheets which correlate with changes in sea level (below).

Zermatt glacier

A 19th century engraving illustrates the growing interest in glaciers, their structures and mechanisms of erosion. The trains of medial and lateral moraine rock debris are clearly shown on the surface, while a falling boulder depicts how further debris is accumulated through rock falls from the precipitous flanking rock walls.

movement. Glacial erratics – boulders carried far from their original sources by the ice and dumped wherever the ice finally melted – were described and traced back to source. Eventually some very distinctive and unique rock types were found which could be used as reliable indicators. For instance, boulders of an unusual kind of granite called ailsacraigite after the island of Ailsa Craig (technically a riebeckite microgranite) off the western coast of Scotland were found to have been carried as far south as Dublin on the east coast of Ireland. Scandinavian metamorphic rocks were found scattered over the North German plain and down to the northeast coast of England.

Gradually all the classic erosive features of upland glaciation and the largely depositional ones of lowland glaciation were revealed – the cirques (also known as corries, or cwms); the U-shaped glacial valleys with their "hanging" tributaries; the roches moutonnées; and the depositional features such as lateral and terminal moraine, drumlins, kettle holes, eskers, outwash sands, and gravels. Agassiz emigrated to America in 1846 and carried on his great glacial conversion campaign. All the same features were to be found around the Great Lakes and everywhere to the north, and on a much grander scale than seen in Britain and even continental Europe. Evidently, high latitudes of the northern hemisphere had suffered a major climate change in the not-too-distant geological past. The question was, why?

The idea that major climate change such as an ice age might be connected to the elliptical orbit of the Earth around the Sun was first promoted by a French mathematician, Joseph Adhémar (1797–1862); however, it was a remarkable Scot James Croll (1821–90), who really developed this concept scientifically.

As geologists investigated the land-based record of glaciation, they realized that there was definite evidence for alternations of glacial and warmer interglacials. For instance, the discovery of a hippopotamus skeleton buried in glacial deposits by the banks of the River Cam in Cambridge, England, provided striking evidence that some interglacials must have been even warmer than present climates at these latitudes. Equally, the discovery of fossils of woolly mammoths in the same area of England showed just how drastically the climate had changed. The problems lay in determining how

One man who became converted to the glacial ideas of Charpentier and others was the Swiss palaeontologist and expert on fossil fish, Louis Agassiz (1807–73).

Armed with this evidence of glacial processes and mechanisms, Agassiz showed British geologists such as William Buckland (1784–1856) and Charles Lyell how to spot the signs of past glaciation in the landscape. Soon the scientific literature was filled with descriptions of ice-scratched and grooved rock surfaces, with the alignment of the marks indicating the trend of ice

James Croll (1821–90)

James Croll was a brilliant and largely self-taught scientist. He worked as a millwright, carpenter, shopkeeper, and insurance salesman before settling in Glasgow, aged 36, as a janitor in the Andersonian College, a job which gave him access to an excellent library and time to read as much science as he wanted.

As a result of these studies, Croll combined measurements of the ellipticity of the Earth's orbit and regular changes of the equinoxes to calculate how the poles could regularly suffer cooling sufficient to generate periodical ice ages, with warmer interglacial periods in between. He also discussed the role and impact of ocean currents on climate.

His ideas, first published in 1864, had a big impact among the scientific world, and soon elevated Croll into a professorship and fellowship of the Royal Society, one of the oldest surviving scientific societies in the world, which had been founded in London in 1660.

many climate swings there had been and over what interval of time.

The task of trying to match sequences of fossiliferous interglacial deposits, mainly lakebed sediments, between different regions of northern Eurasia was difficult enough, let alone trying to match them with North American sequences. The problems of dating, even when radioisotope dating was available, were also enormous. Each glacial phase had tended to wipe the slate clean by removing many previous glacial-related deposits. Most of those that survive belong to the last glacial phase. Persistent searching over the decades, however, has managed to uncover some older ones, and correlation has been greatly helped by the discovery that the best records of climate change during the Quaternary ice ages actually lie at the bottom of the oceans and locked up in the polar ice sheets.

Glacial flow
Minor glaciers, flowing down from the sites of origin in high bowl-shaped cirques, coalesce with main valley glaciers like tributary rivers. Lateral moraines from the side glaciers are amalgamated to form medial moraine in the main flow. Evidence showed that the glaciers had been much more extensive in the past.

modern elephants. These particular features have been independently verified thanks to the portraits of mammoths provided by our ancestors some 30,000 years ago.

The frozen cadavers of the permafrost are often so well preserved that the soft tissue, gut contents, hair, and DNA can be recovered. Despite recent hopes that mammoths might be resurrected by cloning processes using ancient DNA and living elephant mother surrogates, the chances of viable offspring being produced are negligible, and in any case they would be chimeras, mostly elephant and only partly mammoth. Nevertheless, the recovery of such ancient DNA from cold climates is in itself very interesting and shows that

the modern elephant and mammoth evolutionary lineages diverged from one another some five million years ago. The recovery and analysis of ancient DNA are proving invaluable for resolving other problems as well.

It was not just the famous and familiar beasts of high latitudes that became extinct towards the end of the Quaternary ice ages. The lower latitude faunas of the Americas and Australasia were equally as badly hit. Unique localities such as the tar pits of Rancho La Brea, in today's downtown Los Angeles, California, have provided wonderful insights into life well beyond the ice between 40,000 and 10,000 years ago. Excavation of this site over many decades has uncovered some one and a half million bones, weighing 100 tonnes (98 tons),

Baby mammoth
The remarkably well preserved body of an emaciated 18 month old baby mammoth was found in the frozen permafrost ground of a Siberian gold mine in 1977. Nicknamed Dima, it was found to be 40,000 years old by radiocarbon dating.

Frozen Mammoth
In 1901 an expedition from
St Petersburg recovered as
much as it could of the
frozen remains of a large bull
mammoth which died around
30,000 years ago when it was
about 35 years old. The entire
skeleton, much of the skin,
tissue, and hair, apart from
that of the head, which had
been scavenged by foxes, was
recovered and reconstructed
in St Petersburg Zoological
Museum, where it can be
seen today.

and two and a half million invertebrate fossils. The
animals range from giant ground sloths and sabre-
toothed cats to beetles. La Brea was another natural
trap like the sinkhole found at Hot Springs, this time
filled with sticky tar. The herbivorous animals were
attracted by plants, mired in the tar, and eventually
died. Their carcasses attracted predators of all sizes from
big cats to vultures, and many of these also became
trapped. What was bad luck for them has been good
fortune for scientists, who have been able to reconstruct
a detailed picture of the life of the region. One of the
interesting aspects was the inclusion of animals from
South America.

Life in the southern hemisphere was just as
interesting, with some unique faunas and floras
occupying Australia and the innumerable offshore
islands of the Pacific. New Zealand is particularly
intriguing because it separated from Australia at least 80
million years ago and "drifted" into the Pacific with a
"cargo" of plant and animal inhabitants that continued
to live in a time warp. New Zealand's South Island was
severely glaciated because of its high latitude and the
development of a range of high-altitude alpine
mountains. The cold climate eliminated many of the
eucalyptus trees, but left numerous ancient kinds of
conifers called podocarps and araucarias which have
changed little over the past 190 million years. One of
the latter is the kauri, which holds the world record for
volume of timber in a single tree – the largest recorded
kauri had a girth of 23.43m (76.87ft).

In New Zealand, there has been no significant
climate change over the past 1000 years. Its unique
animals include ancient kinds of frogs, the tuatara
lizard, giant earthworms, crickets, and carnivorous snails,
along with an amazing diversity of birds. Some 245 bird
species, mostly unique, occupied the islands until the
first arrival of humans only 1000 years ago. They
included many flightless species and some giants such
as the moa (*Dinornis novaezealandiae*) which weighed
around 98kg (216lb). Some 40 of those species are now
extinct thanks to the human introduction of rats,
hunting, and agriculture.

Back in Africa, *Homo erectus* gave rise to further populations of migrants which moved into Europe and parts of Asia. Generally recognized as species such as *Homo antecessor*, *Homo heidelbergensis*, and *Homo neanderthalensis*, these relatives had bigger brains and made somewhat more advanced stone tools. But again it was back in Africa around 200,000 years ago that another population arose with more modern human biological characteristics and developed the greater cultural skills for which they are recognized as members of our species *Homo sapiens*. They spread throughout Africa and eventually splintered into a number of populations whose descendants still can be recognized in Africa. Recent discoveries from southern Africa show that, by 90,000 years ago, these people were using sophisticated tools made from bone, were burying their dead with ceremony, and had begun to use graphic symbols.

By this time, some of them had also made their way north out of Africa to the Middle East, where they encountered the incumbent Neanderthal people. Around 60,000 years ago, our African ancestors spread throughout central Asia and got as far as Australia by around 50,000 years ago, perhaps encountering some surviving *Homo erectus* relatives on the way. In western Europe, modern humans and Neanderthals came into close proximity and overlapped for about 10,000 years. It was thought that the Neanderthals may well have been the ancestors of the original Europeans, but ancient DNA evidence discounts this. It has been discovered that the Neanderthals possessed a few genetic signatures which distinguish them as a separate interbreeding species. None of these genetic signatures turns up in the modern European gene pool.

The extinction of the Neanderthals is just like that of the other large Quaternary mammals closely coincident on the arrival of modern humans. That is not to say that there was blood on the ground from widespread conflict. It is more likely that the modern humans outhunted the smaller bands of Neanderthals, whose population gradually dropped below sustainable numbers, just as is happening today to many of the remaining large mammals, such as the rhinoceros and tiger, as a result of hunting, poaching, or the destruction of their habitats by modern man.

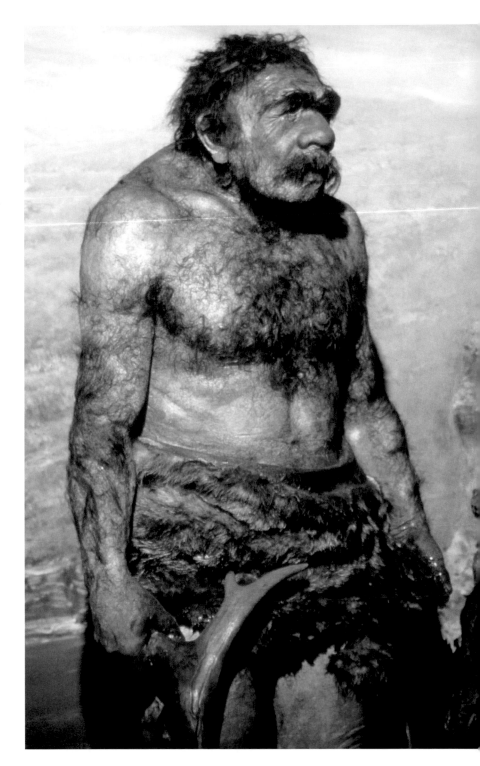

Neanderthal talking point (left)
The most complete skeleton of a Neanderthal is that of an adult male, 60,000 years old, found at Kebara, Israel, which preserves evidence that they had the power of speech.

Neanderthal man (above)
Powerfully built and big brained.

3
Life's third age

Apart from scattered Quaternary age surface deposits which provide the introduction to Earth history, the first layers of this geological story, recount the events of life's third age – the Tertiary as it was originally called. We now divide its history into two periods – the Neogene (1.8–23.8 million years ago) and the Paleogene (23.8–65 million years ago). Together, they are sometimes referred to as the Age of Mammals, although it was just as much the age of flies, flowers, songbirds, modern bony fish and primates. As we shall see, life's third age began abruptly with a bang.

Because these young "Tertiary" deposits generally lie close to the surface, especially in the Mediterranean region, they play an important role in the history of geology. Many of these rocks and their contained fossils were uplifted from the surrounding seas by earth movements associated with the building of the European Alps. This accident of geological history was a lucky one because it meant that many of their fossils were not greatly different from life forms found today.

	3800 MA The first evidence of chemical life on the planet	
PRECAMBRIAN EON		
ERA	4600 MA HADEAN	3800 MA ARCHEAN

CENOZOIC ERA		
PERIOD	65 MA PALEOGENE	

65 MA End Cretaceous extinction	56 MA Diversification and radiation of mammals	41 MA *Eosimias*, the earliest anthropoid
		40 MA Opening of North Atlantic, and volcanism
		38 MA Break-up of Australia and Antarctica
		35 MA Impacts at Chesapeake, USA, and Popagai, Russia

Among these Tertiary fossils, the remains of shellfish, clams, snails, and sea-urchins, as well as sharks' teeth were all clearly recognisable when they were discovered by geologists. The similarities to modern life forms helped these scientists to recognize the fossils as the remains of past life.

The Mediterranean region is still subject to reminders of Earth's dynamism. Earthquakes and volcanic eruptions are common and can easily be linked to particular rock products such as lavas and geological phenomena such as faults. It took centuries for such phenomena to be traced to fundamental driving mechanisms within the earth. Geological wonders such as the Giant's Causeway and Fingal's Cave were celebrated by artists (for example the German composer Felix Mendelssohn, 1809–47) for hundreds of years, but only recently has the Paleogene outpouring of these lavas been linked to the opening of the North Atlantic Ocean, which occurred as the ocean floor spread through the dynamic of plate tectonics.

Fossil stag beetle

Stag beetles are very rare as fossils, let alone specimens as well preserved as this one which still retains some iridescence on its carapace. Comparable forms are today only found in the tropics of Southeast Asia but this one was found at the 49 million year-old Messel World Heritage Site in Germany.

1200 MA The first multi-celled organisms date from the middle of the Proterozoic period

610 MA The first large marine animals appear

PHANEROZOIC EON

2500 MA PROTEROZOIC

545 MA PALEOZOIC 248 MA MESOZOIC 65 MA CENOZOIC

TODAY

23.8 MA NEOGENE

1.8 MA QUATERNARY

20 MA India collides with southern Asia; building of Himalayas

11 MA Late Miocene climate cooling; spread of grasslands

6–8 MA Common ancestor of chimps and humans

3.6 MA Laetoli bipedal footprints

3–3.8 MA *Australopithecus afarensis* (Lucy)

2.4–3 MA *Australopithecus africanus*

2.5 MA Panama freeway and fauna exchange

2.65 MA Oldest stone tools

A world dominated by grasses, insects, and mammals

The terrestrial world of Cenozoic times came to be dominated by familiar life forms ranging from flowering plants (angiosperms) to songbirds (neognaths) and mammals. Likewise, the marine world came to be dominated by sea-dwelling mammals from whales (cetaceans) to seals (phocids), along with a myriad of modern bony fish (teleosts) and the more archaic but still eminently successful sharks. The Cenozoic explosive radiation of all these groups (except the sharks) is one of the most remarkable evolutionary features of the past 65 million years and created the modern biological world.

Most Cenozoic life forms can fairly easily be placed in their general categories, such as cats or whales within the mammals, or grasses within the flowering plants. In detail, however, almost all the species are entirely extinct ones, and the further back we go into Cenozoic

times, the less familiar they become. Exotic-looking groups appear in the fossil record and can be difficult for the nonexpert to identify, even at the class level. Animals such as the diprotodontids of Australia, which were cow-sized marsupial mammals, are unfamiliar or indeed unknown. As we shall see, however, the evolutionary roots of these groups extend deeper into the Mesozoic era, when the living world looked very different.

At the base of the food chain on land were the plants, and the Cenozoic plants of the world's landscapes were not so very different from that of today, at least in general terms. Global vegetation zones were broadly similar, although much of the era had warmer climates than the present. Consequently, the tropical, subtropical, and boreal zones were expanded, and the cooler polar and subpolar zones contracted. For instance, in northwestern Europe, palm trees flourished where there are now woodlands of deciduous oaks, beech, and birch — that was, until the climate began to cool around 11 million years ago in late Miocene times.

The dominant Cenozoic trees included both recently evolved flowering ones and more ancient plant groups such as the conifers. Some isolated regions still carried other plant survivors from Mesozoic times such as the cycads, which had been dominant forms, but went into decline in competition with the flowering plants. The great innovation in the plant world that appeared during the Cenozoic occurred when one group of flowering plants evolved into the grasses (Graminae). It is hard to overestimate the importance of this innovation that helped to transform life. An essential component of the success of the flowering plants was what is called the co-evolution of pollinating insects, which was often very specific. Seed- and fruit-eating animals such as birds, bats, and our ancient human relatives also helped the distribution of many plants.

By opening a series of windows, we can look into some of the changes in past life revealed by the fossil record. While the African "Eden" was hosting the

Messel pit plant
This large (23cm (9in) long) fossil leaf with its distinctive venation is typical of the widely distributed aroid plants belonging to the family Araceae, which today grow as lianas or epiphytes, mostly in tropical and subtropical regions.

PERIOD | PALEOGENE | NEOGENE | QUATERNARY

EPOCH | 65 MA PALEOCENE | 34.8 MA EOCENE | 33.7 MA OLIGOCENE | 23.8 MA MIOCENE | 5.3 MA PLIOCENE | 1.8 MA PLEISTOCENE | 10 KA HOLOCENE

evolution of our human ancestors, North America was occupied by a surprising variety of mammals that one might not have expected to have lived on that continent. In late Cenozoic times (the Pliocene, between 5.3 and 1.8 million years ago), with cooling climates, the older forests covering much of the continent were replaced by extensive grasslands. The secret of the success of the grasses lies in their ability to survive and indeed thrive while constantly being cropped by animals, whether domesticated or natural grazers. Unlike many other plants, leaf tip removal does not damage growth because leaf growth originates lower down the stem. Also, many grasses can reproduce from underground runners, so that they are able to survive even wildfire or severe cropping.

The expansion of grassland as the forests diminished into patchy woodland promoted the expansion and survival of many kinds of grazing mammals, along with the more established browsers such as the elephant-related mastodons. The horses (equids) first evolved in North America around 55 million years ago as small, dog-sized, woodland-dwelling animals; they diversified into many different kinds, increased in size, and became much more fleet-footed to deal with life out on the dangerous open grasslands. However, they, along with so many other medium- to large-sized mammals, became

Snake

This beautifully preserved snake *Palaeopython* is 2m (6ft 6in) long, and was found in the 49-million-year-old Eocene age Messel oil shales, a World Heritage Site near Frankfurt in Germany. The strata preserve a remarkably complete view of an ancient community of plants and animals, from pollen and leaves to insects and fish, birds still with feathers, and primitive mammals with hair.

Primitive horse

No bigger than a large dog, the primitive horse *Propaleotherium hassiacum* is just one of two species found among some 70 specimens excavated from the Eocene-age oil shales of the Messel World Heritage Site in Germany. It has four hooved toes on each of the front feet and three on each of the back. Some of these tapir-like forest-dwelling mammals have been found with their stomach contents of plant-leaf material.

extinct in the Americas during Quaternary times, only to be later reintroduced by Spanish colonists.

There was also a variety of deerlike animals which included some true deer (cervids) and camel relatives. Inevitably, all these plant-eaters attracted the attention of predatory carnivores, including big cats and some hyena-like extinct animals known as borophagines. The

A successful migrant

The continued success of the Virginia opossum (*Didelphis virginiana*) is largely due to its relatively small size (similar to a domestic cat) and its adaptability. It is omnivorous and will scavenge for a wide range of food, can climb well using its prehensile tail, and has plenty of teeth (50) for defence. The downside is that, despite its hair, it is not very tolerant of the cold.

latter had massive, bone-crunching jaws and were probably carrion feeders, operating in groups to drive off the top predators from their kills.

When North and South America were intermittently linked by land, there was an interchange of animals between the two continents. With the break-up of the Pangean supercontinent, South America, like Australia, inherited a primitive mammal fauna dominated by marsupials. Some of these strange creatures took the opportunity to migrate to North America, but today are represented by just one species, the Virginia opossum (*Didelphis virginiana*), which is still remarkably successful. The migration of more modern mammals from the north into South America sounded the death knell for most of the marsupials, who had not, on the whole, had to deal with fast-moving top carnivores such as the big cats. Some of the more modern mammals, such as the giant ground sloths that were abundant in South America, even returned to North America.

Around 30 million years ago (early Oligocene times), the North American "game parks" would have looked much stranger, with a number of large, now extinct mammals grazing and browsing their way across the landscapes. The grasslands were beginning to spread, but there was still plenty of woodlands, which promoted the evolution of a variety of browsers of different sizes that could reach to different heights into the bushes and trees for their food. The brontotheres such as *Brontotherium* were huge, tapir-like animals with rhinoceros-like horns, standing 2.5m (8ft) high at the shoulder. There were also hornless rhinoceroses such as the 4m (13ft) long *Metamynodon* and a number of primitive horse relatives such as *Mesohippus*. In addition, there were some ferocious-looking scavengers such as the hog-like *Archaeotherium* and the bizarre predatory creodont *Hyaenodon*, which may have hunted in packs like the modern hyena.

At the opposite side of the Earth, the last vestige of the Gondwanan supercontinent broke up as the two huge continents of Australia and Antarctica parted, with the Southern Ocean growing between them as Australia moved north. The break-up occurred in late Eocene times around 38 million years ago, and Australia carried with it a remarkable biota of plants and animals from Gondwana, a few of which still survive today. This is especially true of the plants, but also of primitive marsupial mammals who were once much more common the world over. Australia's surviving marsupials, including the kangaroo and koala, are familiar enough today, and there are photographs of the last Tasmanian "tiger", the thylacine marsupial which died in captivity in 1933 and was the last known example of this species in existence. Yet Australasia had a much more diverse marsupial fauna in the Cenozoic era.

Thanks to some remarkable limestone deposits in Australia's northwest Queensland around the town of Riversleigh, a great deal is now known about these animals over a considerable time span from about 15 million years ago. At this time, the region was heavily forested with lowland rainforests, lakes, and rivers developed on a limestone terrain that was riddled with caves, sinkholes, and underground water systems.

Unfortunately, very few plant fossils have survived from these extensive forests because not only did the forest floor soils break down most organic material, but also the soils were not preserved as geological strata, but were themselves eroded away. However, in and around the lakes and sinkholes, carbonate-rich sediment accumulated in hollows, along with the bones of any animals which had drowned in the waters. Carbonate mud can quickly and easily harden into limestone at surface pressures and temperatures. In so doing, it encases and preserves any bone material.

It was only in 1983 that the remarkable fossil faunas of Riversleigh were first discovered, but since then many thousands of almost perfectly preserved bones have been recovered by carefully dissolving the limestone rock in weak acid. Many hundreds of new marsupials and other animals such as snakes, crocodiles, and birds have been found. The marsupials range from cow-sized plant-eaters such as *Neohelos* through flesh-eating kangaroos (*Ekaltadeta*) and the cat-sized predator *Priscileo*, to a variety of other kangaroos and tree-climbing possums such as *Burramys* etc.

With hindsight, no matter how successful the marsupials, they were a sideshow. It was the evolution of the placentals that ensured the mammals subsequent global success. The ability to retain and nourish (via a placenta) a developing embryo within its mother's body, means that at birth the foetus can be much more advanced. Interestingly, even the most advanced placental babies, such as those of deer and cattle which can be mobile within hours, are no more advanced initially than those of turtles or crocodiles. The big difference is made by the post-partum behaviour of the placental mother, who continues to feed her offspring with nutritious breast milk, enabling rapid growth and development. The placental "technology" also allows for the evolution of different strategies of reproduction, foetal development, and care. Numerous small and relatively helpless (altricial) offspring can be produced, or a few bigger and more advanced (precocial) offspring. However, an important variant on the latter strategy occurs where there are a few large but relatively helpless offspring because development has been concentrated in the sensory "apparatus" and its control centre the brain, a strategy adopted by our group – the Primates.

Opening and closing oceans

Rift valley

From a satellite, the rifts forming as plate tectonic mechanisms break Africa away from Arabia can easily be seen extending north from the Red Sea along the Gulf of Aqaba towards the Dead Sea, and along the Gulf of Suez to the Mediterranean, alongside the Nile Delta.

A major discovery of science in the 20th century has been that of plate tectonics. Central to the understanding of its driving mechanism has been exploration of the ocean floor and its geological structure. Following World War II and developments in submarine warfare and technology, the onset of the Cold War in the 1950s added an urgency to the mapping of the ocean floor. The realization that atomic-powered submarines could hide within deep-sea valleys made mapping ocean-floor topography an urgent priority.

The production of the first published global map of the ocean floor by Bruce Heezen and Marie Tharp in 1977 was a remarkable work of synthesis. For the first time, a clear view of the major features of the ocean floor was available. As we have seen, ocean-floor topography is dominated by a series of interconnected rocky mountain ridges. Within the Atlantic Ocean, it is evident that the ridge lies midway between the American continents to the west and Africa and Europe to the east, and it generally parallels the coastal configuration on both sides. If the continents are rejoined along the midocean ridge, there is a near perfect fit along the continental shelf margin. Sampling of rocks from the ocean floor and from islands flanking the ridge systems shows a similarity of composition in that they are primarily derived from basaltic magmas. Radiometric dating shows that all the ridge rocks are very young and that they become older further away from the ridge. Clearly some mechanism is producing new volcanic rocks along the ridge crest, then carrying them on either side.

The further away from the ridges, the thicker the blanket of ocean-floor sediment becomes until it completely covers the irregular ridge topography. Eventually the sediment builds up into flat-lying, deep abyssal plains some 3–7km (2–4½ miles) below sea level. Drilling cores into this thin sediment blanket reveals that the oldest sediments are found furthest from the ridges, but nowhere is there any sediment older than about 180 million years. It would seem that some process is at work destroying the older ocean floor. It is also evident that in places the ocean floor plunges into even deeper trenches, so deep that some extraordinary force must be holding or pushing the ocean floor down in these zones.

The oceanic ridge system has a number of peculiar features that are fundamentally different from those seen in terrestrial mountain ranges. In cross-section, the ridge is strongly symmetrical, and its crest is frequently offset by transcurrent breaks known as "transform faults". The summit of the ridge is divided by a central fault-bounded rift valley. Large-scale tensional forces must be pulling the ridge apart to produce such rifts. Yet how can tensional forces elevate a huge mountain chain some kilometres above the ocean floor? The answer is provided by the simple physical principle that heated matter expands. Heat-flow measurements show that the rocks beneath the ridge systems are significantly hotter than normal ocean-floor rocks, and calculations show that this is sufficient to expand the rocks upwards to form the ridge. On either side of the ridge, the rocks are cooler and sink deeper down. By contrast, terrestrial mountains have fold and fault structures that are

PERIOD | PALEOGENE | NEOGENE | QUATERNARY

EPOCH | 65 MA PALEOCENE | 34.8 MA EOCENE | 33.7 MA OLIGOCENE | 23.8 MA MIOCENE | 5.3 MA PLIOCENE | 1.8 MA PLEISTOCENE | 10 KA HOLOCENE

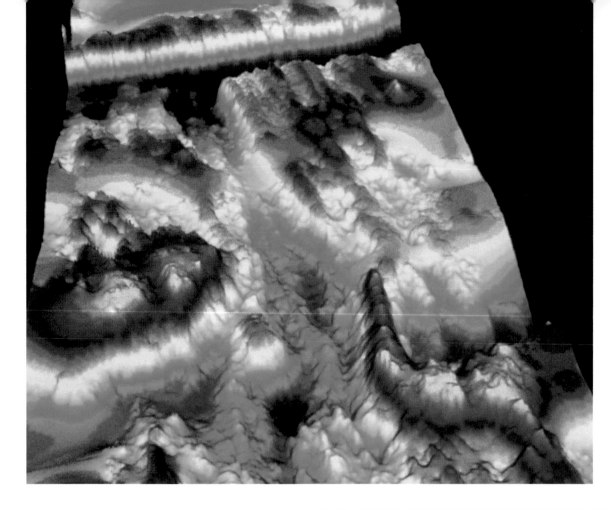

Mid-ocean ridge (left)
Scanned by sonar from a
ship, the Mid-Atlantic ocean-
spreading ridge runs from
top to bottom of this image
with its deep central, fault-
bounded rift valley coloured
blue, and the flanking lava
ridges yellow to red,
according to elevation.
The whole ridge structure is
cut by two transform fault
valleys running from left to
right, one at the top of the
picture, the other at
the bottom.

essentially the products of large-scale compression and
thickening of crustal rocks.

Detailed imaging within the ocean ridge has
revealed numerous conelets and linear fissures mark the
rift floor, just like volcanic rift valleys on land. Expansion
and arching of the ridge rocks generate tension in the
crest and the formation of linear cracks (faults). Partial
melting of the hot rocks below the crest leads to the
upward migration of the melt, which pours out along
the fissures as lava. The lava cools, becomes brittle, and
is in turn cracked open, with new lava pouring out and
so on – a double conveyor-belt system carrying older
lava away on either side of the ridge.

The reality of this process of the ocean floor
spreading away from the ridges was beautifully
confirmed in early 1963 when the Canadian scientist
Lawrence Morley saw a new map of symmetrically
patterned magnetic anomalies on either side of the Juan
de Fuca ridge in the northeast Pacific. It was his "Eureka!"
moment because he immediately realized that it
supported the theory of ocean-floor spreading and heat
convection within the Earth's mantle layer. Like a
barcode, these anomalies of ocean-floor lavas recorded
reversals in the Earth's magnetic field over time. The fact

that the patterns were mirror images on either side of
the ridge and grew older away from the ridge proves
that, over millions of years, the ocean-floor rocks are
literally pushed away from the ridge by new eruptions.
Morley's insight was so innovative that he was unable to
get it published; however, later the same year, two British
scientists, Fred Vine and Drummond Matthews, came up
with the same idea independently and managed to
publish. Convection-driven ocean-floor spreading was the
driving force behind the generation of new ocean floor.

Black smoker (above)
Hot rocks close to the surface
below mid-ocean spreading
ridges heat seawater that has
penetrated the ridge lavas and
recirculates it back to the
surface through vents. Minerals
scavenged from the rocks are
precipitated forming "smokers"
and are fed on by extremophile
bacteria that survive
independently of solar energy.

Building and destroying mountains

Tibetan plateau

Elevated 4–5km (2½ to 3 miles) above sealevel, the Tibetan Plateau, seen here near Lhasa, is the largest high plateau in the world, rising well above the average height of the continents. Coloured Spaceborne Imaging Radar helps reveal the geology of this hilly terrain – granite is orange and brown and older sedimentary and volcanic rock is blue.

Terrestrial mountains are very different from those which rise from the ocean floor. With the average elevation of the land being only a few hundred metres above sea level, a mountain can be defined as any mass of rock elevated significantly above the surrounding landscape. Mountains range from individual volcanic peaks such as Mount Fuji in Japan or Mount St Helens in the northwest United States to ranges which extend across continents and hemispheres such as the Andes of South America. Mountains can vary from sculpted glacial horns such as the Matterhorn in the Swiss Alps to flat-topped elevated mountain plateaux such as South Africa's Table Mountain, as well as the Tibetan plateau and the Altiplano of the Bolivian Andes, with the latter two both rising to more than 5000m (16,400ft).

The major mountains of the world tend to occur within extensive linear ranges which can have very complicated geological histories and structures, such as the Rocky Mountains of North America or the ancient Caledonian mountain belt. The latter stretches from the Scandinavian Arctic Circle, through the northwest of the British Isles and Ireland, and now, since the opening of the Atlantic continues, to Newfoundland and New England.

Compared with oceanic mountain ranges, those of the continents are much more diverse in origin, age, structure, and rock composition. Some mountain ranges contain very ancient rocks that are thousands of million of years old. But these are unusual – most major mountain ranges such as the Himalayas, Andes, and Rocky Mountains are relatively young structures (a few tens of millions of years old). Nevertheless, they are older and very different in origin from ranges found on the ocean-floor.

Typically, land-based mountain or orogenic belts contain folded and faulted rocks produced by large-scale compression. The same forces also metamorphose rocks at depth, turning muds to slates or schists, and limestones to marbles. Rock melts, known as magmas, form intrusions ranging from vast granite plutons with volumes of many cubic kilometres to thin sheet or wall-like dykes and sills. The latter are generally associated with large volcanic centres and the extrusion of lava and other volcanic products at the surface. Such volcanic centres originate from rising plumes of heat deep within the mantle, which dome and fracture the continental crust with vast outpourings of lava. Some 40 million years ago an event of this kind led to the development of the Brito-Icelandic province and the rifting open of the North Atlantic. Similar processes are still at work in northeast Africa and up into the Red Sea, forming the Great Rift Valley, with volcanoes such as Mount Kilimanjaro and flanking mountain ranges such as the Aberdare Range.

Geophysical surveys confirm that the "roots" of young mountains are much deeper – up to 80km (50 miles) deep – than the elevated portion is high. Powerful compressive forces, resulting from the collision of crustal plates, thicken the crust by folding and faulting. Very large-scale faulting can also thicken the crust by underthrusting large slabs of rock over hundreds of kilometres at low angles. Such processes have probably

PERIOD	PALEOGENE			NEOGENE		QUATERNARY	
EPOCH	65 MA PALEOCENE	34.8 MA EOCENE	33.7 MA OLIGOCENE	23.8 MA MIOCENE	5.3 MA PLIOCENE	1.8 MA PLEISTOCENE	10 KA HOLOCENE

been responsible for the 5000m (16,400ft) elevation of the Tibetan plateau, north of the Himalayas.

Plate tectonics can explain much of the compressive force that generates linear mountain belts. The formation of the Himalayas is fairly simple compared with more complex orogenic zones such as the Rockies. As part of the Gondwanan supercontinent break-up, the Indian subcontinent split away from Africa around 80 million years ago. The subcontinent was carried northwards as the Indian Ocean grew in size behind it and the Tethys Ocean was subducted in front of it. The subduction process generated volcanoes, and sediments were scraped from the ocean floor, diced, and wedged onto the continental edge, eventually to be caught up in the mountain-building process. Around 40 million years ago, India collided with the southern margin of Asia, and so began the formation of the great Himalayan mountain range.

Where two oceanic plates collide, the subduction of one of them generates volcanicity and the formation of mountainous volcanic island arcs such as the Philippine

islands. Large island arc masses resist subduction with the result that they can become part of a growing orogenic zone along with other kinds of plate fragments known as "exotic terranes". There is evidence that the coastal ranges of North America include crustal fragments or exotic terranes that have been transported great distances by plate movements before colliding with the North American plate and being incorporated into the larger orogenic belt.

Detailed studies of the rocks and structures (folds, faults, and other features) of mountain belts both ancient and modern have shown that their development can be exceedingly complex. Now, thanks to the emergence of plate tectonic theory, many of the processes of mountain building can be understood, and the "deep" history of plate movements over geological time reconstructed. The development of mountain ranges now separated by oceans, such as those of the eastern seaboard of North America and northwestern Europe, can be better explained. But many gaps in our knowledge remain especially in relation to the early history of the Earth.

Mississippi Delta
Major rivers such as the Mississippi-Missouri system carry very large loads of sediment, much of which reaches the sea and is deposited in huge deltas which build out well beyond the coastline onto the continental shelf. Such sedimentary environments accumulate large quantites of organic debris, which will form hydrocarbons in the form of coal, gas, and oil in the geological future.

Evolution

The 1861 discovery in Germany of *Archaeopteryx*, still the oldest bird fossil we have, provided the first good proof from the geological record of the Darwin/Wallace theory of evolution. Initially, the theory was jointly published joint in 1858 under the title *On the tendency of species to form varieties; and on the perpetuation of varieties and species by means of natural selection*. When, in 1859, Darwin expanded his ideas in *On the Origin of Species by Means of Natural Selection, or the Preservation of Favoured Races in the Struggle for Life*, he actually said very little about fossils. He did, however, say a lot about the deficiencies of the fossil record. He knew only too well that there were eminent scientists with a much greater expertise on fossils, such as his ex-Cambridge tutor Adam Sedgewick, who were antipathetic to the whole idea of evolution. Darwin had been forewarned about the kind of hostile reception the theory was likely to receive because, in 1844, the anonymous *Vestiges of Creation*, outlining general ideas about evolution, had been severely criticized.

Darwin's five-year voyage on the *Beagle* from December 1831 to October 1836 had exposed him to the amazing diversity of life in the tropics and the remarkable adaptations of life on ocean islands, which was particularly well demonstrated by the finches of the Galapagos Islands. By 1838, he had read the Reverend Thomas Malthus's seminal *Essay on the Principle of Population* (originally published in 1798) and was formulating the beginnings of a theory of evolution. As secretary of the Geological Society in London, Darwin was at the hub of a buzzing scientific revolution and network of the foremost scientists of the age. He heard all the eminent fossil experts of the time arguing over new fossil discoveries, the nature of the fossil record, and its interpretation.

Was Charles Lyell right in thinking that fossil representatives of most major groups of organisms could be found even in some of the oldest fossiliferous rocks, or was there evidence for some progression in life from the most primitive to more advanced forms? Even by the 1840s, the jury was still out on this question. In 1842, the anatomist Richard Owen (1804–92) realized that new fossil discoveries showd a very advanced group of reptiles, which he named as the dinosaurs, had become extinct. If the theory of evolution were right, how could such a successful and well-adapted group of animals as the dinosaurs have become extinct? Darwin did not want to become mired in such arguments and stuck to biological evidence to support his theory.

One of the major historic debates of biology was over the fixity of species. The majority of naturalists thought that species could not change in any lasting way, although the famous French biologist Lamarck (1744–1829) had argued that change, promoted by usage, could be passed from generation to generation. For this he was roundly attacked by other biologists, especially his compatriot Georges Cuvier.

Darwin was no supporter of Lamarck's approach, but rather looked to the practical experiences of breeding domestic animals for his ideas. Cattle, sheep, horses, dogs, and some birds (pigeons and fowl) had all been transformed within a few hundred years and a few tens of generations by selective breeding. Could similar selection processes produce such change over time in nature? And what about hybridization? It was well known that crossing a horse and an ass produced a hybrid but sterile mule, which seemed to support the idea of the fixity of species. But there was also evidence that some animals, such as cats or cattle, could be successfully hybridized. Darwin argued that speciation could happen in nature through isolation of populations and adaptation to different environments and biological circumstances. The problem, however, was that the mechanism of such change was not known in Darwin's day.

Richard Owen was one of Darwin's contemporaries and an ambitious man. Owen used his position as keeper of the natural history section of the British Museum to spend the nation's money on important and interesting new fossils and biological specimens. So, when he heard

Ichthyosaur (right)
A well preserved skeleton of a marine ichthyosaur reptile, found in the 19th century, had its tail straightened by collectors at the time to look better.

Lyme Regis (far right)
Since the early 19th century fossils have been collected from the Jurassic limestones and shales which make up the cliffs around Lyme Regis in southern England. Our modern-day "Mary Anning" ought to be wearing a hard hat because rocks regularly tumble down such cliffs.

in a circular motion to "fly" through the water. Many kinds of these remarkable animals are now known, and they range in time through the Mesozoic from Triassic to late Cretaceous times. They all died out, however, before the end Cretaceous extinction event. In fact, the mosasaurs seem to have replaced the ichthyosaurs as top predators before they, too, became extinct.

Even by the mid-19th century, it was realized that some contemporary groups of sea creatures did die outright at the end of Cretaceous times, notably the molluscan ammonites and their distant relatives the belemnites. The ammonites became especially important because they were so diverse and abundant as fossils. In the 1830s and 1840s, scientists such as the Frenchman Alcide d'Orbigny (1802–57), the Germans Friedrich Quenstedt (1809–89) and Albert Oppel (1831–65), and William Williamson (1816–95) in Britain recognized that fossils could be used to make fine subdivisions of strata. Linear successions of rapidly changing but related species could be used to identify and match successions of strata over large distances. The development of a biozonal scheme based on ammonites was first developed by Quenstedt and Oppel, and is still used in a refined form today. Individual biozones represent periods of time that are as little as one or two million years.

Mary Anning

Mary Anning (1799–1847) was a remarkable woman whose contribution to the scientific study of fossils is only now being fully appreciated. She was one of the few survivors of 10 children born to Mary and Richard Anning of Lyme Regis, a small fishing village in Dorset, England. Impoverished, the family even spent time in the workhouse (1811–15), so out of sheer necessity took to scavenging the local beaches for anything saleable. These were the days of Jane Austen, and the growing moneyed middle class took to visiting quaint villages by the sea. The Anning family sold fossils to these first tourists.

Mary and her older brother Joseph found their first ichthyosaur skull in 1811, and a year later they recovered its body skeleton. The specimen still exists in London's Natural History Museum, but for many years there was no acknowledgement of who had found it. Mary went on to find more important specimens of extinct reptiles, all of which she had to sell. Her surviving letters show that she was making acute observations about the nature of the fossils, and she enjoyed arguing with the experts.

By the late 1830s, however, the fashion for fossil collecting was in decline. Fortunately, the gentlemen of science who had benefited from Mary's finds persuaded the prime minister, Lord Melbourne, to grant her an annual pension of £25 in 1838, but she was probably already ill with the breast cancer which eventually killed her, only five years after her mother's death.

The discovery of the dinosaurs

Today, the common conception of the Mesozoic era is fixated on the dinosaurs of Jurassic Park, as described by the author Michael Crichton and portrayed by Steven Spielberg in his films. *Jurassic Park* is a great story based on the intriguing idea that it was possible to resurrect dinosaurs from the dead through a bit of clever genetic engineering. First, find a mosquito, perfectly preserved in amber and, assuming it has been feeding on dinosaur blood, extract some blood cells and their molecular (DNA) prescription for that particular dinosaur. Next, cut the dinosaur DNA into a reptile or bird egg, so it replaces that of the parent and, hey presto! a dinosaur clone. Unfortunately or fortunately, depending upon your point of view, it is not quite that easy, especially as, despite the reports, no original DNA has ever been extracted and amplified from amber-embedded organisms. DNA is a very fragile molecule that needs very special conditions for its preservation.

Only 200 years ago, it was the extinct marine reptiles, the ichthyosaurs and plesiosaurs, that first grabbed the public imagination because theirs were the first complete skeletons to be recovered and reconstructed. The main reason for this was that most of the Jurassic and Cretaceous strata of Western Europe, where the pioneering work was being carried out, is made up of marine sediments, limestone, mudstone, and sandstone full of the great diversity of fossil organisms which inhabited these warm, shallow waters. There was a scattering of islands, however, and land was never all that far away, although few terrestrial sediments are preserved in this region. Nevertheless, some fragmentary remains of animals and plants which occupied these territories were washed into the nearby seas, where they were preserved to be discovered by geologists in the first few decades of the 19th century.

With hindsight, it is now realized that dinosaur bones were first found many centuries ago in China and even in Europe, but nobody understood what they were. The finds that really set the ball rolling were those of a 25cm (10in) long jawbone with curved, blade-shaped teeth from Jurassic rocks in Oxfordshire, England and some strange leaf-shaped teeth found in Cretaceous strata in Sussex, England. It was William Buckland (1784–1856) who became interested in the Oxford teeth and the question of the kind of animal to which they might have belonged. Clearly, they were those of a large predator and, as they were all of a similar shape, it was more likely that they belonged to a reptile rather than a mammal, which has teeth of various shapes. Buckland named the creature *Megalosaurus*, meaning "giant lizard". But what kind of reptile was it? The only models available at the time were crocodiles and lizards.

The Sussex teeth were radically different and were studied by a local doctor, Gideon Mantell (1790–1852), who was determined to try to make his name as a scientist. Mantell was sure the teeth belonged to some new kind of animal, but did not know what kind, and so he canvassed the opinion of the French anatomist Georges Cuvier. Cuvier at first dismissed them as belonging to some kind of plant-eating mammal. Mantell persisted in his searches and eventually obtained a large slab of rock with a jumble of bones from the same strata from which the teeth had come. There were limb bones and lots of back bones, but unfortunately no skull. Again, Mantell had to struggle to find a model for his reconstruction. Luckily he was shown the pickled body of a marine plant-eating iguana lizard which had recently arrived in London from the Caribbean, and he saw to his amazement that its teeth were remarkably like his fossil ones, only much smaller. Scaling up from the metre-long iguana, Mantell concluded that his fossil animal also had to be a quadrupedal reptile at least 10m (30ft) long and a plant eater, and he named it *Iguanodon* in 1825.

Meanwhile, the brilliant young anatomist Richard Owen (1804–92) was a rising star and one of a new breed of professional museum-based scientists in Britain. He was constantly on the lookout for any opportunity to grab the scientific headlines and get one over

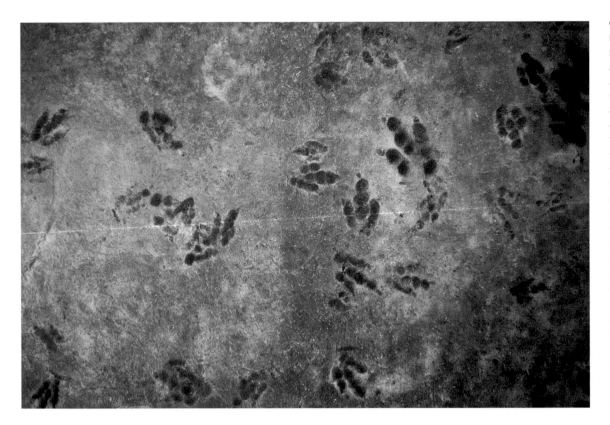

Giant bird footprints

In the mid 19th century, Edward Hitchcock, an American professor at Amherst College in Massachusetts made a detailed collection and study of fossil footprints from Triassic terrestrial strata in the region. Most were three-toed bipedal prints similar to those made by birds only much larger, and Hitchcock's inevitable conclusion was that they must have been made by giant birds. He named and classified the prints according to their detailed appearance (*Brontozoum* in the upper picture and *Gigandipus* in the lower one). Many of his names are still in use, although his giant birds are now known to be dinosaurs and other extinct reptiles and amphibians.

Dino-drama

Dinosaurs have replaced the mediaeval dragon as an icon for monstrous ferocity and danger. The sheer size of many dinosaurs has provided illustrators with an endless source of inspiration for their imagination. Here, in a late Cretaceous scene, two carnivorous tyrannosaurs are depicted attacking a rhinoceros-like, plant-eating ceratopian. In the background, the 10km- (6.2 mile-) wide Chicxulub impactor is seen approaching the Earth and is about to cause the extinction of all the protagonists.

contemporaries such as Buckland and Mantell. Owen had been commissioned to make a particular study of all the newly discovered problematic reptile fossils which were being uncovered from Mesozoic strata in England. In 1841, he presented his review to the British Association for the Advancement of Science. The following year he published his report, which included some important modifications and additions, including the naming of a whole new taxonomic group of extinct reptiles, and so grabbed the limelight from the unsuspecting Buckland and Mantell. Owen coined the name "dinosaur", meaning "terrible lizard", and in doing so opened a Pandora's box of dinosaur mania that still thrives today – if only Owen had registered the name, his descendants would be exceedingly rich.

Although reconstruction of the dinosaurs was constrained by the available reptile models, Owen

realized that such large animals would not have been able to support their massive bodies with crocodile- or lizard-style limbs. When the Crystal Palace from London's Great Exhibition of 1851 was relocated to Sydenham, South London, Owen was given the perfect opportunity to promote his vision of this extraordinary new group of extinct reptiles. Owen tutored some of Queen Victoria's children and had the ear of Prince Albert, who in turn was enormously influential in promoting science and technology in Victorian Britain.

The world's first theme park devoted to prehistoric life, complete with the first life-size dinosaur models, was created by Owen and the artist Benjamin Waterhouse Hawkins (1807–1889) around the rebuilt Crystal Palace. Owen constructed his scale-covered reptilian dinosaurs as curious hybrids with thick mammal-like, elephantine legs supporting massive bodies and heavy, muscular tails which drooped to the ground. Buckland's *Megalosaurus* and Mantell's *Iguanodon* were both seen as huge, cumbersome quadrupeds.

The question of evolution was very much in the air at the time because of the anonymous publication of the bestseller *Vestiges of Creation* in 1844. Owen was highly critical of the work and thought that the evident success of the dinosaurs as a group followed by their extinction argued against ideas of "fitness" and progression. The reopening of the Crystal Palace in 1854 drew thousands of people and so successful was the exhibition that Waterhouse Hawkins was invited to New York to create a similar one for Central Park.

By the time Waterhouse Hawkins arrived in New York, the image of the dinosaur was already changing. By the end of 1858, the substantial part of a new dinosaur called *Hadrosaurus* had been shown by the anatomist Joseph Leidy (1823–91) to belong to a large, bipedal animal. Waterhouse Hawkins set up a workshop and was modelling *Hadrosaurus* when the entire scheme fell foul of local politics and collapsed. *Hadrosaurus* was further transformed in 1866 by American dinosaur expert Edward Drinker Cope (1840–97) into a lithe, leaping, kangaroo-like animal far removed from Owen's imperial leviathans.

Europe still had something to contribute to the *Iguanodon* story. In 1877, the massive bones of some 31 virtually complete members of the genus were found underground by coal miners at Bernissart, in Belgium. They were reconstructed by Professor Louis Dollo using kangaroo and bird models as giant bipedal plant-eaters up to 9m (30ft) long which could use their muscular tails as a kind of prop in order to reach high into tree canopies. It is now known that their tails were stiff, straight, balancing structures that pivoted the body around the pelvis and above the massive hind legs. Fossil tracks show that they were predominantly quadrupedal, but could perhaps rise up on their hind legs for feeding.

By the 1880s, the great American dinosaur race was on, with teams from different museums and universities competing with one another to find bigger and better skeletons in the American Midwest. Othneil Marsh (1831–99) of Yale made the first reconstruction of a

giant plant-eating sauropod 18m (59ft) long, which he called *Brontosaurus*; however, in doing so, he had cobbled together the remains of two different dinosaurs, including *Camarasaurus*, which is the only name now recognized. Not until 1908 was the first of the awesomely large predatory dinosaurs found in late Cretaceous strata in Montana by an expedition from the New York–based American Museum of Natural History. The 15m- (49ft-) long beast was seen as the king of the tyrants of the dinosaur world and given the name that every child over the age of four now knows – *Tyrannosaurus rex*.

Dollo's dinos
One of the most spectacular dinosaur finds in Europe was that of 31 early Cretaceous age *Iguanodon* skeletons in a Belgian coalmine in 1877. The paleontologist Louis Dollo reconstructed some of these 8m- (26ft-) long skeletons as bipeds with a kangaroo-like stance. We now know that they walked on all fours but could rise up on their hind legs to feed on high vegetation.

Life in the shadow of the dinosaurs

The existence of mammals throughout much of Mesozoic times is often overlooked because the dinosaurs receive all the attention and glory. Certainly it is true that for more than 150 million years most land animals more than 1m (3ft) in length were dinosaurs and their reptile relatives. But in the shadow of these remarkable beasts lurked our remote mammalian ancestors – "wee, sleekit, cow'rin, tim'rous beastie[s]", as the 18th-century Scots poet Robert Burns aptly described their modern descendant, the mouse.

These primitive mammals had good reason to be overawed, cautious, and indeed frightened by the reptile-dominated world around them. Nevertheless, throughout much of the time of the dinosaurs, mammals were evolving and diversifying. They were doing so, however, within severe limits. The dinosaurs were not all large and cumbersome – they ranged from pigeon-sized to the famous gigantic sauropod monsters which were pushing the anatomical limits for life on land. A consequence of this diversity was that the dinosaurs and their reptile relatives occupied most ecological niches, and there was not much ecospace left for any other beings.

The few niches that remained were small, dark and crypt-like, such as caves, hollows in trees, and underground. Survival in such limited circumstances places severe constraints on body size and lifestyle. It demanded specially developed sense organs – in other words, good eyes, ears, nose, and some mechanism for feeling the way, such as whiskers. Adaptations of this kind could easily be used for life in the dark in general, ie being nocturnal, especially as reptiles tend to be less active at night, at least in cool or cold climates.

Inevitably, such a hidden or nocturnal life also has to be one of opportunity and carried out either in something of a hurry or very cautiously. The operation of all the sense organs concentrated in the head, where the animal first meets the environment, requires an enlarged brain compared with that of the reptiles. And brains burn energy, requiring constant refuelling, as does

the maintenance of a high level of body activity without the warmth provided by the sun. Critical adaptations, then, are warm-bloodedness and an insulating coat of hair.

It used to be thought that warm-bloodedness was a unique attribute of mammals and birds. Over recent decades, however, very good arguments have been presented suggesting that some groups of dinosaurs, especially the small- to medium-sized predators, may well have been warm-blooded too. The discovery of their relatedness to the birds makes this even more likely. Recently, it has been discovered that some of these dinosaurs and their reptile relatives (the flying pterosaurs) had body coverings of modified scales, some of which were hair-like and others which were feather. As they were not used for flight, an insulating function is highly likely.

The constant demand for energy requires food supplies that are preferably high in protein. Insects, other invertebrates, and some plant parts such as seeds are ideal foods, but processing them for rapid digestion and energy release requires a set of specialist tools: teeth. Catching and killing insects requires sharp, dagger-like teeth; crunching hard carapaces or shells requires strong, crushing teeth, preferably with some points; and chopping the food into bite-sized pieces requires ridges or blades – all powered by strong jaw muscles. Tooth differentiation such as this is a particular characteristic of the mammals, although again dinosaur diversity was such that groups such as the oviraptorosaurians (eg *Incisivosaurus*) also evolved a degree of tooth differentiation. There were also important groups of non-dinosaur reptiles that began the trend, particularly the cynodonts of Permian and Triassic times.

The other major advance in adaptation seen in the mammals was in their means of reproduction. Reptiles and birds lay eggs with shells, which is a considerable advance on the unprotected eggs of the amphibians. However, either the eggs have to be big enough to allow

the hatchling to emerge and immediately fend for itself or, if the egg is smaller, the parent has to feed the hatchling until it can do so. Both strategies were and are adopted by reptiles, but eggs and hatchlings are good food and attract the attention of predators. Laying enough eggs to ensure survival of at least some of the hatchlings is not very cost effective. Retaining and feeding the developing embryo in the mother's body until it is more mature can give it a good start in life, but it cuts down on the number of offspring and is a considerable drain on the mother's resources.

Nevertheless, there were enough advantages in the method for it to become the main mode of reproduction in mammals. There are, however, surviving primitive mammals that remind us that this was not always so for the group.

The few egg-laying monotremes of Australia such as the platypus and the echidna – as well as the much greater number of marsupials, with their very underdeveloped offspring kept in pouches after birth – provide good evidence that the more advanced placental mode of reproduction was a relatively late development

Mongolian egg-basket

The first well-preserved dinosaur eggs, nests, and hatchling skeletons were found in Mongolia's Gobi Desert by Roy Chapman Andrews' expedition from the American Museum of Natural History in New York in the 1920s. Since then the region's Cretaceous strata have produced a wealth of fossil dinosaurs and early mammals.

in mammals. The problem for the palaeontologist is that fossilization of the skeleton does not record much information about the niceties of reproduction. Indeed, fossilization tends not to preserve much at all about the small primitive mammals that crept around in the dark while the dinosaurs were not looking.

Small mammal skeletons, as with those of birds, are fragile and easily broken down either by scavengers or microbial processes of decay. The only long-lasting remains are the more resistant teeth, which can pass right through the gut of a scavenger. As most birds do not have teeth, their fossil record is even worse than that of small mammals. Fortunately, mammal teeth are not only preservable, but remarkably diversified in their form and generally distinctive even to the species level. From this tooth-dominated fossil record, several primitive and short-lived fossil mammal groups are now recognized, such as the multituberculates, triconodonts etc.

We now know that from mid-Jurassic into mid-Cretaceous times at least five different and mostly short-lived groups of primitive mammals evolved and died out. There were three surviving groups, one of which was multituberculates, which survived until some 35 million years ago before dying out. The other two are the monotremes and the marsupials, both of which have just about hung on into the present. Right at the end of Cretaceous times, however, a new group of placental mammals evolved. Among this new group of placentals were the earliest primates, small lemur-like animals

Mongolian mammal

Like most early mammals, *Zalambdalestes* was shrew-sized and lived a cryptic existence alongside the dinosaurs and other reptiles of late Cretaceous times in Mongolia. From these somewhat unspectacular beginnings all mammals evolved.

Mining the record of life

The popular view of life on earth in Jurassic times 150 million years ago may be seriously distorted. Tens of thousands of fossils recovered over a period of 10 years by German scientists from the Guimarota coalmine in Portugal record a very different picture from that depicted in the film *Jurassic Park*. Sure, there were plenty of dinosaurs about – 15 different species, identified from some 750 fossils – but none was much bigger than a turkey, and they were outnumbered by lots of other reptiles, especially small lizards, turtles, flying reptiles, and crocodiles. Some of the latter were "salties", marine crocodiles in other words, and by far the biggest beasts around, growing to 9m (29ft) long, substantially bigger than the Australian saltwater crocodiles alive today. The scientists also discovered 9,000 or more fragments of frogs and salamanders, not to mention more than 100 teeth belonging to the earliest bird, *Archaeopteryx*, – the only record of the animal outside of one locality in Germany. All in all, more than enough fossils to keep some 20 experts busy for many years.

Most of the fossils are minute teeth and bones, which have had to be laboriously extracted and then hand-picked from the underground coal deposits. Altogether, the fossils and accompanying sedimentary rocks open a unique window on life in a subtropical coastal swamp just as complex as that of the Florida Everglades today. The biggest surprise is the diversity of mammals, with some 25 species (identified from 7,000 teeth and 800 jaws), again outnumbering the dinosaurs. These distant hairy ancestors of ours were all small, shrew-like and none bigger than a hedgehog. Their delicate bones are not normally preserved, so we usually gain a very biased view from the fossil record which preferentially preserves big bones. Thomas Martin, one of the German team, does not exaggerate when he says: "The Guimarota mine is the most important fossil lagerstatte (accumulation of well-preserved fossils) of the world for late Jurassic mammals and other small terrestrial animals."

which lived up in the trees. A new analysis of the rather poor fossil record we have of the early primates suggests that the last comon ancestor of the primates probably lived as long ago as late Cretaceous times, around 81.5 million years ago. This estimate coincides closely with the divergence times estimated by the molecular clock analysis from living primate groups such as the apes, monkeys, lemurs, tarsiers, lorises, and galagos.

Luckily for us, some of these mammals did manage to survive the Cretaceous/Tertiary extinction event. Within 10 million years, they had become so successful that they split into some 15 different mammal groups, ranging from bats to whales. Even so, six of them subsequently became extinct.

Even as far back as the early decades of the 19th century, before the discovery of the dinosaurs, primitive fossil mammals were found in the same Stonesfield slate strata of Jurassic age which yielded some of the first dinosaur remains. The fossils were small jawbones and teeth, and, when Georges Cuvier examined them in Oxford University's Ashmolean Museum in 1818, he thought that they belonged to ancient marsupial mammals. He also recognized, however, that they differed from living mammals by having more molar teeth.

One of the specimens was indeed illustrated, described, and named in 1825 as a fossil marsupial, *Thylacotherium*, by the French palaeontologist Constant Prevost (1787–1856). In 1846, Richard Owen showed that it was not a marsupial, but a primitive placental mammal, and he renamed it *Amphitherium*. Nevertheless, one of the other Stonesfield mammal fossils was a genuine marsupial. Even in 1853, Charles Lyell saw these discoveries as "... fatal to the theory of progressive development, or to the notion of the order of precedence in the creation of animals". Within a decade, Lyell had to eat his words and grudgingly acknowledge the concept of evolution.

Insect-eating ancestors

Henkelotherium is a primitive mammal whose well-preserved remains were found in a late Jurassic-age coalmine in Portugal. A woodland species, somewhat like a tree-shrew, it hunted the abundant insects living in the trees whose decayed remains eventually formed coal deposits.

5
Life's extinction events

Ideas about the nature of change in prehistoric life have vacillated over the centuries between the sudden and the gradual. The classical western and Judeo-Christian view tended towards the sudden and catastrophic. God was wrathful and capable of both creating and destroying life in an instant. The Old Testament saw life created within days, and virtually wiped out by the Noachian Flood.

Early scientific views of the prehistory of the Earth and life tended to build on this catastrophic vision. The succession of sedimentary rock strata with their fossils was thought to result from one or more catastrophic floods. However, by the beginning of the 19th century, naturalists who explored the fossil record found that it extended much deeper into the stratigraphic rock record than previously expected. Furthermore, fossil representatives of the major groups of living organisms continued to be found in older and older strata.

3800 MA First evidence of
chemical life on the planet

PRECAMBRIAN EON

ERA	4600 MA HADEAN	3800 MA ARCHEAN

PALEOZOIC ERA

PERIOD	545 MA CAMBRIAN	495 MA ORDOVICIAN	443 MA SILURIAN	417 MA DEVONIAN	354 MA CARBONIFEROUS

443 MA End Ordovician
extinction, 53% marine genera extinct

450 MA Late Ordovician glaciation

The Scottish geologist Charles Lyell, who exercised a powerful influence on 19th-century scientific thought, even suspected that fossils of all groups would be traced back to some creative moment in the remote past. But he soon had to abandon that position as it became clear that there was some sort of progression from primitive to more advanced life forms through geological time. This latter view was supported by the Darwin/Wallace theory of evolution, which was essentially gradualist and required small changes over a very great deal of time.

At the same time, paleontologists were finding that the distribution of life through geological time was not as smooth as Darwin expected it to be. Even by the 1860s, John Phillips, William Smith's nephew showed that the three great eras of life, the Paleozoic, Mesozoic, and Cenozoic were separated by significant declines in diversity which are now recognized as major extinction events. It now seems that there is both gradual and catastrophic change, the question is, which is the more significant?

Precambrian Ice Age

This mixture of irregularly shaped boulders scattered through finer grained deposits in Varangerfjord, Norway is typical of a fossil till deposit generated by glacial erosion and deposition. The scratches on the rock pavement in the foreground are evidence for glaciation.

2500 MA "Snowball" Earth glaciation	1200 MA The first multi-celled organisms date from the middle of the Proterozoic period	700–580 MA "Snowball" Earth glaciation	610 MA First large marine animals appear	575 MA Abundant Ediacarans	

PHANEROZOIC EON

2500 MA PROTEROZOIC	545 MA PALEOZOIC	248 MA MESOZOIC	65 MA CENOZOIC

MESOZOIC ERA　　　　　　　　　　　　　　　　　　　　　　**CENOZOIC ERA**

MIAN	248 MA TRIASSIC	205 MA JURASSIC	142 MA CRETACEOUS	65 MA PALEOGENE	23.8 MA NEOGENE
	248 MA Permo-Triassic extinction event, globally low sea-levels, 68% marine genera extinct	200 MA Eastern North American plateau basalts	133 MA Parana (South America) and Etendeka (SW Africa) basalts	65 MA Chicxulub impact event, 43% of marine genera extinct	
	250 MA Siberian flood basalts	200 MA opening of Central Atlantic		65 MA Deccan lavas	
		205 MA End Triassic extinction, 45% marine genera extinct	133 MA Opening of South Atlantic		

Danish boundary layer

The Cretaceous/Tertiary boundary with the iridium spike is seen in many localities around the world. Indeed, its global occurrence is evidence of the magnitude of the impact event. Here, for instance, at Stevns Klint on the Danish coast, the boundary is marked by a clay full of fish remains separating uppermost Cretaceous chalk from earliest Tertiary chalk.

species lasts indefinitely, but either dies out or evolves into one or more new species, and this process is an ongoing one. Typically, species only persist for, at most, a few million years. Human-related species such as the Neanderthals (*Homo neanderthalensis*) only survived for some 300,000 years before becoming extinct without evolving into a new species, although at one time it was thought that they may have contributed to the modern European human gene pool. (DNA evidence has subsequently disproved this.) Nevertheless, the extinction of the Neanderthals may well be seen as part of a wider but relatively minor extinction event, associated with the latter part of the Quaternary ice

ages and the destructive spread of modern humans. Although this event has been spread over 50,000 years and more, on the geological time scale, this is still considered a rapid event.

In the last few decades of the 20th century, serious efforts were made to collate and quantify changes in fossil life through time. American palaeontologists Jim Valentine and J. John Sepkoski Jr led the way. The vast, scattered scientific literature, written in many languages over the past century and a half, was scoured for information about the numbers of fossil taxa, their distribution, and their duration throughout geological time. This was attempted at different taxonomic levels to see what patterns emerged, but it was soon realized that there are many problems with this kind of data set. For instance, many taxa are duplicated because different scientists have given them different names. The accuracy with which their stratigraphic ranges are given varies enormously and tends to refer to the nearest known boundary, so that inevitably there is a false clustering at boundary intervals. Certain geological strata, especially the younger ones, have been investigated to a greater extent than older strata, in what is known as the "pull of the recent". Some geographical areas, such as Europe, have been investigated in more detail than other parts of the world. Nevertheless, it was still an extremely useful exercise and drew a picture of the diversification of life over time which was a significant advance on Phillips's 19th-century version.

The distribution curve was based on families of marine creatures because theirs is the best and most consistent record in rock strata. The pattern showed an initial steep increase in diversity from late Precambrian times, accelerating through the Ordovician only to drop dramatically at the end of the period. There was a recovery through the Silurian, but the overall diversity reached a plateau and remained there throughout the rest of the Paleozoic, terminated by another dramatic fall at the end of the Permian, just as Phillips had shown more than 100 years previously. Recovery during the ensuing Triassic was interrupted by another sharp fall at the end of the Triassic, but thereafter there was a steep recovery through the rest of the Mesozoic era, with total diversity finally climbing above that of Paleozoic times during the Jurassic. But then the Cretaceous is terminated by yet another sharp fall, followed by a rapid recovery and another acceleration through early Cenozoic times which has slowed, but still climbs to the present.

Here was graphic evidence of at least four major prehistoric setbacks to the evolution and diversification of life. These seemed to be clear extinction events, but were they real and, if so, what had caused them?

Life's extinction events

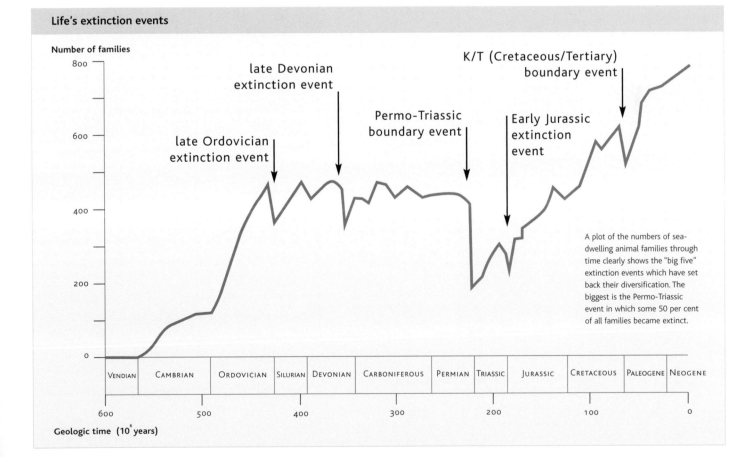

A plot of the numbers of sea-dwelling animal families through time clearly shows the "big five" extinction events which have set back their diversification. The biggest is the Permo-Triassic event in which some 50 per cent of all families became extinct.

Extraterrestrial events

By the 1980s, the extinction event that took place at the end of the Cretaceous period was a well-known phenomenon, in particular the demise of the dinosaurs. Many theories had been advanced to explain it, ranging from virus pandemics to climate change stimulated by very large-scale volcanism. The latter seemed a very real possibility because the one big event known to have occurred at this time was the outpouring of the Deccan plateau basalts in western central India.

Arguments based on why certain fossil groups became extinct and others did not were not helping to clarify what the cause might have been. Scientists from different disciplines were employing their own expertise and techniques to attack such problems. Some of the best sections through the Cretaceous/Cenozoic boundary, known generally as the K/T boundary, are in seabed sediments, especially those found in Europe.

The problem is that, because the Earth is still fortunately a dynamic planet, the surface is constantly being reworked by a variety of geological processes which are remarkably good at removing even the largest natural phenomena such as mountain ranges and large-impact craters. Even so, to have caused such an extinction, the K/T impactor must have been big and have left an even bigger hole, perhaps in the order of a 100km (60 miles) or so wide. As it happened only 65 million years ago, there should be some signs left on land, but what if it had landed in the ocean? With the more rapid reworking of ocean floor material, it was quite possible that the site would never be found because it could have been subducted. And even if it had not, it could only be "spotted" on the deep ocean floor by some indirect means such as a geophysical survey. Such was the interest in the story, however, especially because of the newsworthiness of the dinosaur connection, that lots of scientists were very keen to receive the credit for finding the impact site.

Meanwhile, other geologists were tackling different aspects of the event on land. They soon found that the iridium "spike" anomaly could be picked up in other sections of the K/T boundary around the world. In addition, close examination of the boundary sections turned up another strange sign of the true nature of the event: shocked quartz grains. Tiny, sand-sized particles of the common mineral quartz were found with sets of parallel fracture planes running through them that are not normally seen in this mineral. Experiments showed that they could only be produced by very high-pressure shock waves, even higher than those induced by volcanic explosions. It was also realized that the relative abundance of shocked quartz was likely

The K/T iridium anomaly

During the 1960s, Italian experts on microfossils were working on a particularly good section near the historic Tuscan town of Gubbio. The strata here show a marked change in the kinds of microscopic foraminiferan fossils found immediately below and above the boundary, which is itself marked by a thin non-fossiliferous clay layer. An American geochemist by the name of Walter Alvarez sampled the clay layer as part of an analytical survey and was surprised to find that the clay contained a very rare element called iridium. Although it was only present in very minute quantities (a few parts per billion), iridium is normally much rarer than this in the Earth's crust. In fact, it is only known to occur in the Earth's core and in extraterrestrial rocks such as meteorites left over from the early formation of the Earth.

Walter Alvarez's father happened to be the Nobel Prize–winning physicist Luis Alvarez, so naturally Walter discussed this strange iridium anomaly with his father, who took no time in piecing the puzzle together. The only way the iridium could end up in a deep marine clay was from dust introduced by the impact of a very large body from outer space. Walter knew very well that one of the most famous extinctions to occur at the end of the Cretaceous was that of the dinosaurs. In 1980, Alvarez father and son (along with Frank Asaro and Helen Michel) published their ground-breaking speculation that it was just such an impact event that had caused the extinction of the dinosaurs and so much else. The question was, if there had been such a globally catastrophic event, where was the big hole that would have been caused by it?

to be higher nearer the impact site, and soon mapping of shocked quartz distribution seemed to be pointing towards the general region of the Gulf of Mexico and the Caribbean.

In retrospect, the site of the impact was found in the early 1980s by Mexican geologists doing exploratory drilling for the Mexican state oil company. They even published a short synopsis of their discovery in a major American geological journal, but nobody else realized the significance of what they had found at the time. Then, in 1991, a geophysical survey picked up a whole series of anomalies over the Yucatán peninsula in Mexico and out into the Gulf of Mexico. These showed a large, hidden, multi-ringed structure blanketed by a kilometre of more recent sediments. The diameter of the furthest ring was more than 200km (120 miles), and there was a clear high in the middle of the structure. The site of the K/T

impact had been found and became known as the Chicxulub crater, after the nearest town in the peninsula.

The discovery of the impact site coincident with the K/T boundary extinction event reinvigorated investigation of the whole phenomenon. How could an impact event, however big, in the Gulf of Mexico cause a global extinction affecting life in both the sea and on land?

The amount of iridium found in the boundary spike suggested that the impactor had to be an asteroid or comet some 10km (6 miles) in diameter, which would have been vaporized by the energy of the impact. It was thought that the dust thrown into the atmosphere would have caused a scenario similar to a "nuclear winter", with catastrophic loss of sunlight, atmospheric cooling, and a collapse of plant life at the base of the food chain. It turns out, however, that things were more complicated than that.

Dinosaur extinction

More fancy artwork depicting the K/T apocalypse with a *Tyrannosaurus rex* in the middle of a desert, (an unlikely setting for such a large meat eater) with falling meteorites. Nevertheless, there is the evidence of a big hole in Mexico which shows that a 10–20km- (6–12½ mile-) wide impactor from space did indeed hit the Earth with devastating effect about 65 million years ago at the end of the Cretaceous period and was at least partly responsible for the extinction of a wide range of creatures, including the dinosaurs.

Volcanism and climate change

Deccan traps

Coincident with the 65-million-year-old impact event was the outpouring of incredible volumes of basalt lavas over the Deccan region of western central India. Greenhouse gasses released by the huge scale of the volcanism may well have contributed to rapid climate change and enhanced the magnitude of the extinction of life.

Geological mapping of the remoter parts of the world has revealed vast areas covered with basaltic lavas known as plateau basalts. These basalts are not necessarily linked to individual volcanoes, but are the product of large-scale fissure eruptions, and their nature has been something of a problem until recent decades.

Some of them, such as the Miocene age Columbia River basalts, which may have poured out 170,000km³ (40,785 cubic miles) of basalt, and the Tertiary (end Paleocene age) Igneous Province of the northwest part of the British Isles (Brito-Arctic region), have been known about for a long time. Since the advent of plate tectonic theory and their accurate dating, however, a new picture has emerged for these extraordinary events which have punctuated Earth history. It turns out that a number of these basalts coincide with both major and minor extinction events.

Major ones include the 65-million-year-old Deccan traps of India, estimated to have erupted some 2,000,000km³ (479,825 cubic miles) of lava, which coincides with the end Mesozoic extinction. There are also the 251-million-year-old Siberian traps (between 2–3,000,000km³ (479,825–719,738 cubic miles) of lava which coincide with the end Permian extinction. The eastern North America plateau basalts of the New Jersey region, which may have originally had a volume of some 2,000,000km³ (479,825 cubic miles), coincide with the

smaller end Triassic (Norian) extinction event. Inevitably, such coincidences have led to speculation that there might be a connection between extinction events and large-scale basalt outpourings.

Rising plumes of heat through the mantle (also known as mantle hot spots) promote such outpourings of lava. They cause doming through expansion of the crust, which in turn stretches the uppermost surface layer. The heat induces partial melting of basaltic rocks at the top of the mantle, and these fluid melts rise to the surface, especially along tension cracks opened by the stretching forces acting on cool, brittle surface rocks. Many past mantle plumes have been precursors to continental rifting and the opening of new oceans in eruptions such as those at the end Jurassic age Parana/Etendeka flood basalts of South America and southwest Africa, respectively, which are linked to the opening of the South Atlantic and those of the end Palaeocene Brito-Arctic Province, linked to the opening of the North Atlantic.

The effects of this kind of volcanicity which relate to global extinction revolve mainly around their gaseous products entering the atmosphere, leading to climate change. Sulphur dioxide and carbon dioxide are volumetrically the most important gases. Sulphur dioxide is a greenhouse gas, and its initial effect is to cause global warming; however, it soon reacts with water in the atmosphere to produce sulphate aerosols which absorb the Sun's radiation, leading to subsequent cooling and acid rain in the longer term (lasting up to 10 years). Any volcanic ash in the atmosphere also promotes cooling, but so far modelling of such pyroclastic elements cannot be linked to extinction events. The release of carbon dioxide has a much longer-term effect, but its cumulative effects could well be climatically significant.

The end Permian Siberian traps are unusual in that they contain large volumes of volcanic ash deposits precipitated from the atmosphere. Their abundance is thought to have been responsible for climate cooling, which some experts think was severe enough to cause glaciation over the period of eruption, some 600,000 years. This, in turn, may have been responsible for the major fall in sea level at the end of the Permian. There is as yet no supporting geological evidence, however, for an end Permian glaciation. Furthermore, there is evidence from a number of sections through boundary strata that, on the contrary, there was a rise in sea level at the time. The most persistent signal associated with the end Permian extinction is global warming, to which the Siberian volcanic activity, which was the biggest outpouring of plateau basalts known, could have contributed. However, the role of methane has also been invoked recently, and this does not emanate from the Siberian fissure eruptions.

The Deccan outpouring of basalts at the end of the Cretaceous also very interestingly coincides with an extinction event. The lava pile reached a peak thickness of 2.5km (1.5 miles), and its total outpouring of some 2,000,000km^3 (479,825 cubic miles) of lava is thought to have happened within one or two million years at most. Estimates of the associated outpouring of carbon dioxide amount to some 500,000,000,000km^3 (11,956,379,829 cubic miles). Calculations of global warming associated with these figures, however, suggest that it would only amount to 1–2°C (1.8–3.6°F). By comparison, similarly large outpourings of sulphur dioxide could have led to short-term cooling, which cumulatively could have lasted longer because of the time scale of the total eruption.

Surprisingly, then, it seems that so far it is not clear what exactly the effects of such large-scale volcanic eruptions might have had on life forms and how they would have contributed to the extinction events. For instance, sediments formed in between eruptions are fossiliferous in places, containing remains of freshwater fish and amphibians; however, these show little evidence of any extinction, even though they are right in the middle of the event.

The dinosaur record is even more curious in the Deccan, for there are fossil fragments of dinosaur eggshells which occur above an iridium anomaly which presumably was caused by the impact event. If this is correct, then, right in the middle of the Deccan is evidence that dinosaurs might actually have survived

into the earliest part of Tertiary times. Alternatively, they may have been reworked from older layers.

There is independent evidence for global climate cooling at the end of the Cretaceous period before the impact event, and there are faunal extinctions that precede the impact such as the demise of the rudist bivalves and many benthic forams. The onset of cooling also, however, precedes the eruption of the Deccan traps by four to six million years. The cooling trend continued into the earliest Tertiary time (Paleogene), but was punctuated by a brief reversal with a 500,000-year phase of warming at the end of the Cretaceous. This coincided with a fall and then rise in sea level. The warming phase coincides with the eruption of the Deccan traps, which may indeed have been responsible for it.

Mount St Helens
The 1980 eruption of this volcano, dormant since 1857, in the Cascade Range of Washington State, northwestern USA was another reminder of the Earth's continuing dynamism. On 18 May it erupted, sending up a plume of ash 19km (11.8 miles) high. The eruption killed everything within an area of 180km^2 (69.5sq miles), and spread ash even further. The eruption blew off about 3km^3 (0.7cubic miles) of the volcano, reducing its height from 2,950m (9,678ft) to 2,549m (8,363ft).

Changing sea levels

Geological evidence for significant changes in ancient sea levels has been mounting over the past 200 years. In Britain and Scandinavia, early 19th-century geologists discovered curious stepped features stretching back from present-day coastlines and sometimes reaching tens of metres above present-day sea level. When investigated in detail, these terraces often displayed beach-type sand deposits, including the shells of shoreline creatures, wave-cut notches, and even sea caves and old sea cliffs. Had the sea level fallen since then or had the land risen? Either way what had caused such changes? Were they local or more widespread and perhaps global?

We now know that the question of past sea level change is really quite complex, with several factors involved. Some changes are just regional, such as those of northwest Europe where the land has rebounded since the melting of the Pleistocene ice sheets, which had literally weighed the land down. This isostatic rebound, as it is called, is now a well-understood phenomenon and still continues today. Over and above this process, however, there were also global changes in sea level during the Pleistocene. During glacial phases, so much water was withdrawn from the oceans, cycled through the atmosphere and precipitated as snow to become locked up as glacial ice (on both land and sea) that global sea levels fell significantly. During warm interglacial climate phases, the ice melted and the water was returned to the oceans, causing sea levels to rise again.

The overall changes amounted to as much as 100m (330ft), which might not seem much compared with the depths of the oceans. However, such changes have made enormous differences to the life of the past. The basic reason for this is the nature of offshore submarine topography. The continents extend beyond their immediate coastlines and are fringed by a shallow submarine shelf called, appropriately, the "continental shelf". This slopes gently away at first, but then comes to an end with a marked steepening of the slope at depths of around 200m (660ft), then rapidly descends into the ocean depths of 2,000m (6,600ft) or more. Strictly speaking, the continents end where this break in slope, known as the "continental margin", occurs. The width of the shelf varies enormously depending largely upon the geological environment and the plate tectonic situation.

Where there are converging plate margins, such as around much of the Pacific basin, the shelves are very narrow because subduction of the ocean floor is occurring offshore and is typically marked by a submarine trench. Any sediment shed from the land into the sea is dumped into the trench and has no opportunity to accumulate. In contrast, where there are passive margins such as around much of the Atlantic basin, shelves are much wider and are largely made up of great wedges of sediment from the land. These accumulate in great piles offshore over many millions of years, especially where great river systems, such as the Mississippi–Missouri into the Gulf of Mexico or the Amazon into the Atlantic, discharge into the ocean with huge deltas.

These continental shelf accumulations of sediment are one of the most important geological features of the Earth for a number of reasons. The shelf surfaces are sites where much of the diversity of marine life thrives, especially where reefs develop. The turnover of such huge biotas results in the burial of very large amounts of decaying organic matter, along with their residual hard parts, which are to become the bulk of the future fossil record. Buried organic debris, especially within or around deltas, can lead to the development of hydrocarbons over the long term, providing the circumstances are right. The sediment fringes themselves represent future sedimentary rocks. Such large accumulations are hard to destroy. When the plate configurations and movements change and passive margins become convergent ones with plate collisions, these sediment piles take the brunt of much of the collision forces and are crumpled into new mountain ridges.

The relative shallowness of continental shelf seas means that a fall of 100m (330ft) or so in sea level will expose a significant proportion of the continental shelves around the world and decimate most of the coral reefs and other life forms which normally occupy these areas. On the positive side, for land-living creatures, any such fall would in the long run provide huge new territories to colonize. Perhaps even more importantly, many islands, such as the British Isles and adjacent continents, would become reconnected by land bridges, such as those between Asia and North America, and Asia and Australia. Reconnections such as this have opened freeways for exchange of life, leading to radical changes in dominance and diversity. For instance, sea level changes during the recent Pleistocene ice ages allowed modern humans to migrate into Australia as long as 50,000 or more years ago, but kept modern humans out of the Americas until some 15,000 years ago.

Many decades of effort have been required for geologists to discover what the geological history of global sea level change has actually been. To prove that events have indeed been global rather than regional, evidence from rock strata has to be accurately matched between continents. Sea level curves have now been established for Phanerozoic times, but their accuracy diminishes for earlier periods. Taking present levels as a standard, it appears that sea levels are now at their lowest for most of Phanerozoic time, except perhaps at the end of the Permian. A number of cycles of rising and falling levels have been recognized, with periodicities of tens of millions of years. Over and above this are longer term changes. Tracing back through Cenozoic times, levels generally rise to a peak of nearly 400m (1,300ft)

Raised beach

Scotland, along with Scandinavia, is still slowly rising out of the sea (known as isostatic rebounding) as a result of the removal of the ice load after the Quaternary ice age, which ended around 12,000 years ago. Here in the Western Isles of Scotland an old sandy bay (now grass-covered machair) and surrounding seacliffs now lie a few metres above present-day sea level.

6
Life's early age

The Paleozoic era of ancient life was not investigated in any detail until the 19th century, although some strata were familiar prior to that through their economic use. For instance, outcrops of coal associated with strata of Carboniferous age were exploited in Europe from the late 18th century onwards. Their excavation soon brought numerous fossils to light, especially those of the plants that made up the coal deposits. However, much of lower Paleozoic-age strata formed mountainous uplands in northwestern Europe, having been subjected to folding, faulting, and metamorphism. Consequently, their complicated structure was not subdivided into geological systems and periods until the latter 19th century.

As this mapping progressed and the Cambrian, Silurian, and eventually Ordovician periods were defined, their fossil faunas were discovered and described. Practically all the fossils were new to science, as they belonged to long extinct taxa.

PRECAMBRIAN EON	3800 MA First evidence of chemical life on the planet	
ERA	4600 MA HADEAN	3800 MA ARCHEAN

PALEOZOIC ERA						
PERIOD	545 MA CAMBRIAN	495 MA ORDOVICIAN	443 MA SILURIAN	417 MA DEVONIAN	354 MA CARBONIFEROUS	29
	515 MA Jawless fish	Jawed fish appear	Upright-growing land plants appear	Forests develop; the first four-legged vertebrates appear	Fying insects and amphibians appear	Reptiles appear during the late Carboniferous period
	525 MA Radiation of invertebrates					
	545 MA Many small shelly fossils					

Lifetimes were spent investigating the amazing lower Paleozoic fossils in regions such as Bohemia in central Europe, (described by Joachim Barrande) and New York State (described by James Hall). It soon became evident that while landgoing vertebrates were found in strata as old as the Carboniferous, they were not present in lower Paleozoic strata, although fossil fish were found. Their record extended back into the Silurian, with many strange armoured creatures without bony jaws or teeth. Mostly, the lower Paleozoic strata were filled with

marine fossils; even primitive land plants did not seem to be present until Devonian times. Clearly, some sort of progression in ancient life was gradually emerging, although it was too vague to be of use to Charles Darwin as he marshalled evidence for his theory of evolution. In fact, Darwin went to great pains to avoid referring to the fossil record because he knew it was still too full of gaps and problems. Today fossils provide countless examples which reinforce the Darwin/Wallace arguments for the origin of species.

Fossilized fern

Many fossil plants, such as this *Alethopteris* frond found in late Carboniferous coal-bearing strata from Pennsylvania, USA look superficially like modern ferns. However, they are gymnosperms (seedplants) and some, like *Alethopteris*, whose fronds grew to 7m (23ft) long, were tree-sized.

1200 MA The first multi-celled organisms date from the middle of the Proterozoic period

610 MA The first large marine animals appear

PHANEROZOIC EON

| 2500 MA PROTEROZOIC | | 545 MA PALEOZOIC | 248 MA MESOZOIC | 65 MA CENOZOIC | T O D A Y |

MESOZOIC ERA

CENOZOIC ERA

| MIAN | 248 MA TRIASSIC | 205 MA JURASSIC | 142 MA CRETACEOUS | 65 MA PALEOGENE | 23.8 MA NEOGENE |

The first dinosaurs and mammals appear

Birds and flowering plants appear in the late Jurassic period

124 MA First placental mammals

Primates and songbirds appear

5 MA First hominids appear on earth

Major extinction event

Welsh Silurian strata

Exposed on the Welsh Atlantic coast, these tectonically tilted sandstones and shales contain fossils such as the extinct graptolites, which show them to have been early Silurian sea-bed deposits, now known to date back some 435 million years.

time to investigate God's handiwork. Many of them sincerely felt that it was their duty to uncover the "testament of the rocks" and reveal the details of the formation of the Earth only briefly alluded to in the Genesis account.

Sedgwick had a much more difficult time, as the geological structures he was facing were often complex and fossils rare, so that it was difficult for him to match sequences of strata from one place to another. His was truly pioneering work, and he made slow progress, especially when it came to describing the fossils which were so critical for identifying and correlating the strata.

Nevertheless, by 1834 the two young geologists had met up and concluded that they could identify two distinct and sequential systems of rock strata. Murchison's was the younger, and in 1835 he named it the Silurian rock system after a Romano-British hill tribe, the Silures. Sedgwick's older Cambrian rock system was also named in the same year, after the Roman name for Wales. Murchison noted that the top of "his" Silurian system passed up into the strata of the Old Red Sandstone and that its base passed down into sandstones and shales belonging to Sedgwick's Cambrian. Furthermore, there were still more rocks below the base of the Cambrian, and they became generally referred to as "Precambrian", to replace the old term Primitive.

By 1839, Murchison published a large book entitled *The Silurian System*, in which he described the various subdivisions of the strata and illustrated their features along with the all-important fossils contained within them. In a series of technical papers, Sedgwick described the details of his strata, but omitted to describe and illustrate the fossils. The two friends went on to see how the Old Red Sandstones fitted into the scheme and expanded their work into Devon in southwest England, where these strata are well exposed. By 1839, Murchison and Sedgwick jointly proposed that the succession of slates, sandstones, and limestones in Devon which lay stratigraphically above Silurian strata could be distinguished as another younger system in its own right. The two travelled to the Continent and found similar rocks and fossils in the Rhine and Eifel regions of Germany. They named the new system the Devonian, after the English county, and showed that its

strata passed up into the younger Carboniferous rock system which had been named in 1822 by two other British geologists, the Reverend W. D. Conybeare and W. Phillips. Thus Murchison and Sedgwick could truly claim that the old term "Transition series" was no longer needed.

The partnership was not to last much longer because of a growing dispute over the boundary between the Cambrian and Silurian systems. Murchison expanded the Silurian downwards, claiming that fossils that he had described were to be found in the upper part of the Cambrian, therefore making it Silurian, and that furthermore the origin of life was to be found within the Silurian. Even when some of Sedgwick's supporters described fossils from the Cambrian strata in the 1850s, Murchison tried to claim that they were Silurian in age. When Murchison became director of the British Geological Survey, he was able to ensure that his version was imposed upon all official geological maps. The growing acrimony between the two men led to their complete estrangement, which was only partially resolved in their old age. Murchison, meanwhile, had travelled to Russia and carved out another system of strata above the Carboniferous, which he named the Permian after the city of Perm in the Caucasus.

Creation of the Ordovician

In the 1870s, Charles Lapworth (1842–1920), a Scottish schoolmaster turned geologist, finally resolved the boundary dispute between the Cambrian and Silurian by carving out the Ordovician system between the two. Lapworth had made detailed studies of an extinct group of fossils known as graptolites, which are often abundant in these ancient strata. Combined with growing evidence of fossils being widely distributed throughout Cambrian strata – especially in Bohemia (part of today's Czech Republic) and New York State – Lapworth claimed that three divisions based on the succession of fossils could be recognized. These were the Cambrian (then thought to contain the oldest fossils), Ordovician, and Silurian, which together comprised the Lower Paleozoic. This is now the internationally accepted arrangement. The Burgess Shale of British Columbia (a World Heritage Site) and the Chengjiang strata of China (which has produced some of the most famous fossils we have) are both acknowledged to be part of the Cambrian system, which represents a period of time from around 545–495 million years ago.

Fossil fuels fire the Industrial Revolution

The Industrial Revolution was integral to the building of our modern world, with all its benefits of technology, science, and medicine, and the drawbacks of social upheaval, industrial poverty, and environmental damage. The earliest phase in western Europe began in the late 18th century, continued into the 19th century, and was largely based on geological factors and the ready availability of coal and iron ore in conjunction with navigable waterways, nearby ports, and markets.

All these factors were found in Shropshire, England, where local names such as Ironbridge and Coalbrookdale recall the start of the Industrial Revolution. Local deposits of coal and iron ore led to the establishment of foundries, the cast-iron products of which were carried by river down to the port of Bristol, which exchanged its infamous and fading trade in slaves for the rising trade in iron goods. The availability of coal was fundamental to the success of the enterprise, and the growing need for ever greater supplies fed a frenzy of exploration to find new sources of the "black gold".

Across northwestern Europe, the growth of the Industrial Revolution was based on this availability of coal, which was mostly Carboniferous in age. Coal-bearing strata were found scattered across northern France, Belgium, Germany, central and northern England, South Wales, and central Scotland. Even when the Industrial Revolution hit North America, the early supplies of coal in the east of the continent, scattered from Nova Scotia down to Pennsylvania, were Carboniferous in age (the system is subdivided into the Mississippian and Pennsylvanian in the United States). The problem was that rocks of this age (now known to be 354–290 million years old) only have a limited exposure at the surface; much more of it is hidden and deeply buried beneath younger strata. The question was how to find these valuable buried coal-bearing strata and then how to extract the coal. Huge sums of money were wasted as landowners frantically dug pits and shafts in the hope of finding coal on their land when the simplest of geological surveys would have shown the

search to be pointless. Inevitably, there was a demand for proper geological surveys and mapping, which initially were carried out by surveyors such as William Smith, until state-run geological surveys were established (in 1835 for Britain).

Academic interest in coal and coal-bearing strata was much older. Edward Lhwyd (1660–1709), the distinguished naturalist and keeper of the Ashmolean Museum in Oxford described and illustrated fossil plants and insects from Coal Measure strata in 1699. He thought, however, that they had grown within the rocks from seed derived from the living ferns they resembled, seed that had been washed through crevices into the rocks. By 1804, the German naturalist Ernst von Schlotheim (1764–1832) had shown that the Coal Measure plants of Thuringia in central Germany had a genuine resemblance to the living tree ferns of tropical regions, but that in detail they were different and represented a totally extinct ancient flora.

By 1828, the French botanist Adolphe Brongniart (1801–76) had distinguished four separate phases in the development of plant life, beginning with those primitive fern-like ones of the Upper Paleozoic, followed by the first appearance of conifers, then cycads in the Mesozoic, and finally the flowering plants in the Cenozoic. Brongniart argued that the profusion of giant tree ferns, clubmosses, and horsetails in the Coal Measures indicated that the climate of the time must have been as hot as the tropics are today. The implication was that, as such deposits were found in quite high latitudes, the Earth must have been much hotter in the remote past. His was one of the first analyses of ancient climate, and he was right to connect the Coal Measure plants with tropical climates. It was not until the acceptance of the plate tectonic theory, however, that it was finally realized that climates had not changed that much in the past, but rather the continents themselves had moved.

The growing exploitation of Carboniferous coal gave a great boost to the study of the fossil plants from

which coal is made, but associated with the plants were some intriguing animal fossils. In 1852, Charles Lyell visited North America, landing at Halifax, Nova Scotia, where he was met by John William Dawson (1820–99), a superintendent of education who took a keen interest in geology (he later became professor of geology at McGill University). Together they visited the excellent coastal exposures of Carboniferous coal-bearing strata at Joggins, which are several thousand feet thick. Fossil tree stumps, up to 8m (26ft) high, could still be seen in their original position of growth. The fossil roots were known

as *Stigmaria* and could be seen embedded in clayey soils with the trunks growing up through coal seams. These were in turn interbedded with shales and sandstones. The distinctively patterned trunks were generally found separated from the roots and had been given a different name, *Sigillaria* – although, in 1846, the botanist Edward Binney (1812–81) demonstrated that they were related to the same plant. Nevertheless, here was unequivocal proof of the association. Also, Lyell and Dawson noted that the interior of the trunks was filled with sediment with just an outer cylinder of the "bark" converted to

The Black Country

Extensive exploitation of coal in the Industrial Revolution soon changed rural landscapes into heavily polluted "black country" in the manufacturing regions of Europe and North America in the 19th century. The rest of the world soon followed. Within a hundred years we have depleted coal which took millions of years to form.

Coal detail

A section of coal, seen under a microscope at high magnification, shows part of a plant (orange in colour) which has been flattened and buried in a matrix of organic plant debris. The details of coal structure were first worked out by the paleobotanist Marie Stopes, (1880–1958) who is perhaps better known as a pioneer of birth control.

coal. Clearly, when the trees had died in the swamps in which they grew, the interior had rotted away and later been filled with sediment.

Thinking that other kinds of fossils might be preserved within the tree's sediment infill, Dawson and Lyell dug several trees out of the cliffs. On breaking open the sandstone infill, they found fragments of fossil fern fronds (*Sigillaria* and *Calamites*), bits of charcoal, and some bones which even they could see were not fish bones as they expected. They then found some jaws and teeth that clearly belonged to a small tetrapod, which they assumed to be an amphibian such as *Labyrinthodon*. Lyell became very excited by the find. As Dawson recalled, "His thoughts ran rapidly over all the strange circumstances of the burial of the animal, its geological age, and its possible relations to reptiles and other animals, and he enlarged enthusiastically on these points." Noting the surprise of the local resident who was assisting them, "he turned to me and whispered, 'The man will think us mad if I run on in this way.'"

Lyell and Dawson described their new tetrapod as *Dendrerpeton*, thinking that it was a reptile, but it was later shown to be an amphibian. Nevertheless, Dawson

revisited the locality in 1859 and found another small tetrapod about 20cm (8in) long, which he called *Hylonomus* and which was then the oldest known reptile. One of their possible interpretations of the find was that the animals had fallen into the hollow trunks and had not been able to escape; however, recent research has revealed a more dramatic story. The presence of the charcoal shows that wildfire was common in these early tropical forests, sparked off by lightning strikes during storms. The animals almost certainly used the tree trunks for shelter from predators and other adverse conditions, but were killed by a forest fire while hiding there.

The fossil charcoal from such sites is botanically of great interest, as it can preserve remarkable detail of the anatomy, down to the cellular level, of the plant tissue from which it is made. Also, recent investigation of the relative density of stomata – the leaf pores plants use to regulate their "breathing" (gas exchange) – provides a proxy measure of atmospheric conditions and the climates of the time. Basically, when carbon dioxide levels are high, the plants produce fewer stomata, and vice versa.

In late Devonian times, the tree-sized plants of the first forests had relatively high numbers of stomata because carbon dioxide levels were high and climates were globally warm. During the early part of the Carboniferous, carbon dioxide levels started to fall dramatically and climates cooled, switching from a greenhouse to an icehouse state. Sea levels were high, and vast areas of the continents were flooded with shallow seas in which corals flourished and limestones were deposited. Climates became even cooler, more so than those of the present day, and sea levels began falling, creating vast areas of swamps in low-lying continental interiors and around the receding coasts. In tropical regions, these were the environments in which the coal measures developed.

The density of stomata on the leaves of Coal Measure plants becomes much higher, showing that atmospheric carbon dioxide levels were very low and oxygen levels were high (about 35 per cent compared with 22 per cent today). Sea levels frequently fluctuated as they fell, so in coastal regions there were constant changes between shallow marine and shallow swampy conditions, which built up the vast thicknesses of coal measures with their alternations of clays, coals, shales, sandstones, and occasional limestones. The deepening "icehouse" state set off glaciation in high latitudes that persisted into Permian times, although in tropical regions the coal measure forests also continued to flourish. Some of China's vast coal deposits, (producing over 1,000 million tonnes per year), are of Permian age. However, the end Permian extinction saw the biggest change in global vegetation with the loss of the tropical giant clubmosses and high-latitude glossopterids. Of 19 seedplant families (gymnosperms), only three survive into the Triassic – there are almost no early Triassic coal deposits.

Around Edinburgh, the capital of Scotland, limestones of lower Carboniferous (Mississippian) age record some of the critical changes in the life forms that were evolving at this time. Here, the limestones are unusual because they are freshwater and were deposited in lakes around the active volcanoes of the region. The mineral-rich soils promoted a lush vegetation of clubmosses (lycopsids), horsetails (sphenopsids), ferns (pteridosperms) etc. The rotting leaf litter provided home and food for early arthropods such as mites, millipedes, scorpions, and the extinct eurypterids (up to 3m (10ft) long). Crustaceans, shellfish, and fish were abundant in the lakes and rivers. Most interesting perhaps has been the discovery by Stan Wood, a Scots professional collector, of several early skeletons belonging to some amphibious tetrapods.

Varying between about 20cm (8in) and 50cm (20in) long, these little salamander-like animals are very important to our understanding of the evolution of the first egg-laying (true amniote) reptiles. Detailed examination has shown that *Balanerpeton* is clearly an amphibian, as is *Westlothiana*, which was at first thought to be the most primitive reptile ever found; however, *Silvanerpeton* and *Eldeceeon* are both anthracosaurs, as is *Hylonomus* from Joggins in Nova Scotia. Dated at around 338 million years old, these Scottish anthracosaurs are some of the earliest and most primitive reptiles known. Technically they are referred to as reptiliomorphs because their true amniote condition cannot be proved, but they do show significant advances on the amphibian structure.

Tree stumps
Coastal cliffs at Joggins, Nova Scotia, Canada expose Carboniferous-age strata with coal seams and the fossil casts of trees still in their life position. When Charles Lyell and John William Dawson examined some of these fossil trees in the 1850s they found the fossil skeletons of small lizard-like vertebrates inside some of the hollow stumps.

Ancestor of the vertebrates

Among living vertebrate-related animals exists a fascinating little creature, the lancelet, technically known as *Branchiostoma* (it used to be called (*Amphioxus*). It is some 4cm (1³⁄₄in) long, laterally flattened like a slender leaf, and pointed at both ends, so it can be difficult to tell the head from the tail at first. Indeed, the lancelet can swim both forwards and backwards, and lives partly buried, tail first, in sea-bed sands. It feeds by filtering tiny organic particles from the water. Water is sucked in through its mouth, passes over and through a barred sieve-like structure in the throat region, and travels out of the body. This way, the lancelet obtains both oxygen and food, which are removed from the water by tissues covering the sieve structure. It is possible that some of the fossil agnathans were similar filter feeders. Otherwise, there is not much to the head of the lancelet except some photoreceptor cells which act as crude "eyes" and sensory tentacles surrounding the "mouth".

What is particularly interesting about the lancelet is its basic anatomy. There is a flexible stiffening rod called a notochord which runs from the very tip of the head to the end of the tail, encased in two side-by-side and parallel segmented series of muscle blocks. When these muscle blocks contract in waves passing from the tail forwards, the body bends from side to side and moves forwards through the water, the same basic mechanism as seen in most fish. Reversing the contractions sends the animal backwards. Above the notochord lies the dorsal nerve cord. Study of the embryological development of agnathans and true fish shows that the notochord is the precursor to the vertebrate backbone. So this elongate axial stiffening rod is a core developmental preadaptation for the vertebrate body form and mode of life.

The critical feature that differentiates the chordates from the vertebrates is seen in the development of the neural crest cells within the embryo. In the vertebrates, some of these cells break away to form structures such as the eyes, skull support structures (skull), and head muscles. Although the lancelet does not have a true neural crest, it does have cells in a similar position which express similar genes.

Burgess Shale (far right)
High on a mountainside in British Columbia, Canada, the World Heritage Site of the Burgess Shale preserves one of the most diverse and remarkable of early fossil faunas of mid-Cambrian age, some 505 million years old. Like somewhat older Chinese strata, these marine deposits preserve not only a wide range of invertebrate shellfish but also some of the earliest chordates.

Before the fish (below)
The living lancelet *Branchiostoma* is a primitive but still successful fish-like marine animal that grows to 7.5cm (3in) long and has a laterally compressed, tapered body, within which are the basic structures of vertebrate organization such as a flexible rod (the notochord), a dorsal nerve cord, and a sieve-like structure in the throat region. Lancelets are chordates and are remarkably similar to some of the earliest fossil chordates.

about the nerves, blood vessels, brain structure, and gill form which confirms their affinities with the lampreys, which still exist today, in particular.

Some of these extinct agnathan groups were not armoured and looked much more like small "fish" (up to 20cm (8in) long) with laterally flattened bodies covered in small scales, but scales which are structurally and compositionally different from the scales of bony fish. These agnathans swam like most modern fish do – by throwing their bodies into lateral waves using muscle blocks which are serially repeated along the length of the body. The tail has both upper and lower lobes, but these are often of different sizes, which would have pitched the animals either head down or head up, suggesting that they were either water-surface or sediment-surface feeders. Lateral fins, often present as continuous lateral fin folds, helped to stabilize forward motion for normal swimming. The gill openings were numerous (eight to 13), the mouth was circular and right at the front of the head, and the eyes were often quite large. Clearly, these were active and possibly quite fast-swimming animals. Some of them had skins without scales, but our record of these forms is extremely poor. Consequently, not much is known about the important transition from jawless to jawed fish that occurred sometime from the end of the Cambrian and the early Ordovician period. But we can go another step backwards in time.

When we look around at some other primitive animals still alive today, it turns out that there are a few groups (such as the sea squirts and pterobranchs) which have a notochord, but only in the larval stage. More interesting from the evolutionary point of view are the extinct conodonts. These were very successful and abundant marine creatures which evolved in late Cambrian times and became extinct at the end of Triassic times. They were generally small (4cm (1³⁄₄in) or so long), but one giant group (40–80cm (16–32in) long) did evolve in the cold waters of late Ordovician times. Only recently has any fossil information about their body form been found – most of their fossil remains consist of tiny barbed bars, barely visible to the naked eye. Made of calcium phosphate, they look like tiny but sharply pointed teeth, and indeed they were first described as such by the Russian palaeontologist

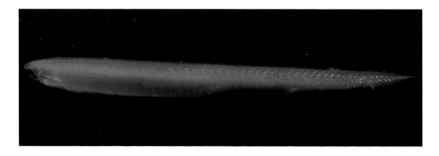

7
Life's beginnings

Our view of the prehistory of life is still very biased towards what
has happened over the last 545 million years since the beginning of
Cambrian times and the appearance of abundant fossils in the rock
record. We now suspect that life probably began soon after the first
primitive atmosphere and oceans had formed on Earth, around 4,200
million years ago. However, the whole business had to start again
because a catastrophic late bombardment of the Earth by meteorites
around 3,900 million years ago is thought to have led to a complete
meltdown and destruction of any initial atmosphere, oceans or life.
Nevertheless, recent discoveries and analyses of what are known
as chemical fossils – hydrocarbon nodules – suggest that by 3,800
million years ago life had probably become re-established. The early
atmosphere and oceans were very different in their chemistry from
those of today and only those single-celled micro-organisms that
could tolerate the conditions, especially the lack of oxygen,
would have evolved.

		3800 MA first evidence of chemical life on the planet	
	PRECAMBRIAN EON		
ERA	4600 MA HADEAN	3800 MA ARCHEAN	

	HADEAN ERA	ARCHEAN ERA	
periods not yet defined	4600 MA	3800 MA	
	4500 MA Impact dislodges moon-sized mass from the Earth	4200 MA Atmosphere and oceans formed	3900 MA Bombardment destroys early atmosphere and oceans	3460 MA First stromatolites

Scientists are actively investigating similar extreme environments today where temperatures are either very high or low and where the water is either very acid or very alkaline, to see what kinds of microbial life can tolerate such conditions and how they do it. The aim is to try and gain some insight into the kind of life forms which might have first evolved.

In recent decades scientists have also tried to replicate the chemical and physical conditions of the early Earth to see if they could generate life from natural combinations of organic chemicals and physical circumstance which might have "sparked" the inherent energy that characterizes living from inanimate matter. Although in some ways successful, these were still very naive and crude experiments but they did force a re-examination of what was known about the basic biology of life, which was very little. The result was a spectacular boom in fundamental research into the workings of the most primitive cells, their chemistry, and molecular biology.

Ancient sediments

Although metamorphosed by heat and pressure, folds and faults, sedimentary lamination is still visible in these late Precambrian rocks. It was from such evidence that 19th-century geologists realized that sedimentation processes seemed to extend back in time indefinitely, beyond the base of the Cambrian.

1200 MA The first multi-celled organisms date from the middle of the Proterozoic period

610 MA The first large marine animals appear

PHANEROZOIC EON

2500 MA PROTEROZOIC

545 MA PALEOZOIC | 248 MA MESOZOIC | 65 MA CENOZOIC

PROTEROZOIC ERA

2500 MA

700 MA First oxygen produced by photosynthesizing microbes

2300 MA First snowball Earth

1200 MA First multicellular algae

720 MA Snowball Earth

580 MA First Ediacarans; Doushanto fossil embryos

555 MA First shelled organisms

The discovery of the Precambrian

Suilven, Scotland
Glacially eroded strata of late Precambrian horizontally bedded sediments (Proterozoic, around 980 million years old) lie on more ancient, highly metamorphosed Precambrian rocks (Archean, dating back to 2,800 million years ago) of the surrounding landscape.

By 1835, when Adam Sedgwick first defined the Cambrian system of rock strata in Wales, the foundations for the division of geological time, based on sedimentary strata and their contained fossils, were literally laid down in tablets of stone. Sedgwick and his geological colleagues were well aware that the testimony of the rocks did not begin with the base of the Cambrian, although they did believe that the story

of life did. In the early decades of the 19th century, the story of life seemed to stretch back to a beginning with the first fossils found in Cambrian strata. Beyond that, there were still plenty of "deeper" sedimentary rock strata which were clearly pre-Cambrian; however, they seemed to be devoid of life and were consequently called Azoic (meaning "without life"). In many places, even older but highly deformed and metamorphosed

strata could be seen, often intruded by igneous rocks – the Primitive rocks of Werner's and other early classifications.

By the mid-19th century, great tracts of apparently Precambrian metamorphic rocks were being found in Scotland and on a much greater scale in North America, where the Canadian Shield was turning out to include a truly vast area of such rocks. Some of these looked as if they had been bedded sedimentary strata before they were metamorphosed. There were, however, problems with trying to unravel Precambrian geological history. As there was no independent method of dating the rocks and no fossils, all that could be done was to try to establish their relative stratigraphic position, which was generally very difficult in highly deformed rocks. Yet, as was being shown in the Scottish Highlands, it was possible to do this. The problem was that intense metamorphism destroys most fossils, so that it could not be assumed that all metamorphic rocks were Precambrian – and, indeed, they are not. In fact, it was to turn out that quite a lot of the Scottish metamorphic rocks are lower Paleozoic in age.

We now know that nearly all the major continents, eg. Laurentia (North America and Greenland), and Amazonia (northern South America), have significant areas of ancient Precambrian exposed at the surface, with younger rocks and often mountain belts wrapped around them. Many of these Precambrian rocks are highly deformed, but not all, and many contain economic mineral resources of global importance, such as the banded iron formations of Hamersley, Western Australia; Isua, western Greenland; Namibia and Transvaal, southern Africa; Siberia; Brazil; and Lake Superior, Labrador, and Minnesota in North America. Their ages range from 3,850 to 700 million years old, with the bulk being between 2,750 and 1,800 million years old. The Hamersley reserves are estimated to be around 27 million tonnes with more than 55 per cent iron in the ore.

Darwin suspected that there should be fossils of primitive life forms to be found in Precambrian rocks. The Scots-Canadian geologist J W Dawson (1829–1899), who had shown Lyell the Joggins fossil trees, made the first breakthrough in 1865. Dawson found some organic-looking remains in Precambrian limestones exposed

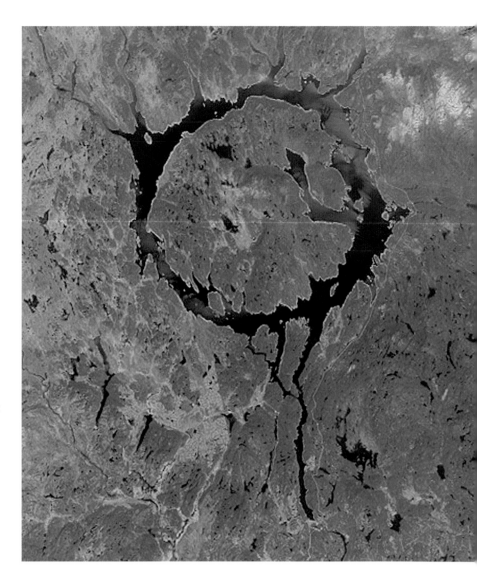

along the Ottawa River near Montreal, Canada and described them as the fossil of a giant, unicellular organism which he called *Eozoon canadense*, meaning "dawn animal from Canada". Dawson's reputation was such that his fossil was generally accepted as the first genuine evidence for life extending back into the Precambrian.

Although it subsequently turned out that Dawson's *Eozoon* was not a fossil at all, but an inorganic mineral growth, the conceptual breakthrough had been made. Geologists now expected to find Precambrian fossils if they looked hard enough in the least deformed of these most ancient sedimentary rocks, and it was not long before they did. By the end of the 1880s, a rising star of American palaeontology, Charles Walcott (1850–1927),

Canadian Shield

At 70km (43½ miles) in diameter, the 220-million-year-old Manicougan impact crater, is still clearly visible indenting the much more ancient Precambrian strata of the Canadian shield. Deep glacial erosion by Quaternary ice sheets has scoured the rocky landscape and infilled the remaining circular depression with melt water.

Eozoon

This cut and polished section through a Precambrian rock from the Canadian Shield was thought to represent the oldest known traces of fossil life and was first described by JW Dawson in 1865. Although its bulging laminae do look organic, more mineralogical detailed analysis proved otherwise.

had found some strange, organic-looking structures in Precambrian strata at the bottom of the Grand Canyon. By 1891, Walcott was convinced that since "life in [Precambrian] seas was large and varied and there can be little, if any doubt that it is only a question of search and favourable conditions to discover it".

By the end of the 19th century, enough was known about the processes of fossilization for it to be realized that it is very selective and generally only preserves organisms with hard parts. As primitive life would lack the necessary hard parts to fossilize and also be very small, if not actually microscopic, the chances of any such organisms being preserved within rocks seemed very slim. Soft-bodied organisms such as jellyfish are very rarely fossilized anywhere in the fossil record, and then only in special circumstances. Microscopic soft-bodied organisms stand virtually no chance of being preserved unless by some process of secondary mineralization.

It was not until the 1950s that such fossils were found – the debunking of Dawson's *Eozoon* had set back the unrewarding task of looking for Precambrian life. Again it was in Canada that the real breakthrough was made, when Stanley Tyler (1906–63), an American economic geologist, found some stunningly well-preserved microfossils in ancient Precambrian Gunflint

ironstones in Ontario which are exposed along the shores of Lake Superior. The ironstones occur as bedded sedimentary ores with layers of chert, some of which are formed in distinctive mound-like structures, similar to those from the Grand Canyon described by Walcott as *Cryptozoon*. At the time, they had no idea of the age in years of the rock, but they did know that it lay deep within the Canadian Precambrian succession. It has now been confirmed as truly ancient, with a radiometric date of 2,100 million years.

It turns out in retrospect that, around the same time, a Russian scientist, Boris Timofeev (1916–82), was discovering and describing similar microorganisms from Precambrian strata in the Urals. As his discoveries were published in Russian journals and this was the period of the Cold War, his pioneering work did not become known about in the West for another decade or so. With the Gunflint discovery, however, it was realized that, when certain special conditions of preservation prevail, microscopic cells that are unprotected by shells or skeletons will fossilize, and so the race began to find just how far back into Precambrian times life extended. The additional breakthrough that helped the whole enterprise enormously was the advent of radiometric dating in the 1950s, again aided by wartime technology associated with work on nuclear fission undertaken by the Manhattan Project. The discovery that Precambrian time amounted to 4,000 million years, some eight times longer than the Phanerozoic with its relatively abundant fossil record, was something of a shock, but it also provided an enormous *terra incognita* for research. Being able to develop some chronology and date rocks more exactly were pivotal to this.

We now know that there are chemical traces of life found in rocks from Greenland as old as 3,850 million years. There is probably a much better chance of preserving such chemical signs of early life than there is of body fossils. However, the nature of such early fossils is hotly debated, and the desire to be the discoverer of the oldest fossils on Earth is as strong as ever it was since the stratigraphic empire-building days of Sir Roderick Murchison.

At the other end of the Precambrian, large organisms had evolved by 600 million years ago, although nobody is entirely sure about what kind of organisms they actually were. They are known as the Ediacarans, after the Australian locality where well-preserved fossils of them were first found. In addition, we now know that the race towards building protective shells also began around 560 million years ago, just before the beginning of Cambrian times.

The whole span of Precambrian time is over eight times longer than the Phanerozoic, which is that part of prehistory we are most familiar with, the last 545 million years. Technically, it has been superceded by major divisions from the most ancient, the Hadean (4,600–3,800 million years ago), the Archean (3,800–2,500 years ago) and the Proterozoic (2,500–545 million years ago). Already, the Proterozoic is subdivided into the Paleo-, Meso- and Neoproterozoic, and as more is learned about the details of Precambrian history, no doubt further subdivisions will be introduced.

Ediacaran homeland

The Pound Quartzite, a late-Precambrian marine sandstone, in the Ediacaran Hills of the Flinders Range, Southern Australia was found by Australian geologist Reginald Sprigg to contain well preserved fossils. They represent a variety of soft-bodied organisms which were thought to be related to jellyfish.

8
Earth's beginnings

Speculation about Earth's beginnings is probably as ancient as the gift of speech – and that may well predate our species, *Homo sapiens* – but it is unlikely that we will ever know the whole answer. Certainly the first signs that can be construed as astronomical speculation are associated with rock art generated by our kind some tens of thousands of years ago. Since then observers from many different cultures over the last few millennia have gone to great lengths to calculate the movements of planetary bodies and have achieved remarkable insights about the position of Earth in relation to the Sun and the other planets. However, it is only since the invention of the astronomical telescope and the advent of intellectual freedom to speculate about Earth's origins unfettered by religous considerations that modern understanding has developed, especially over the last 200 years or so.

PRECAMBRIAN EON

ERA	4600 MA HADEAN	3800 MA ARCHEAN

periods not yet defined	HADEAN ERA 4600 MA

4500 MA Impact dislodges moon-sized mass from the Earth

4400 MA Oldest minerals (zircons); differentiation of Earth's core

4200 MA Atmosphere and oceans formed

The contribution from the rock materials of the Earth to the debate about its origins is much more recent. The dynamic nature of the Earth has meant that the first formed rocks on Earth have either been eroded away or deeply buried and are thus unavailable for sampling. Even those that have been brought to the surface by subsequent earth movements have mostly been highly deformed and altered from their original state by heat and pressure, a process know as metamorphism, which also tends to destroy any fossil remains within them.

Much has depended upon the ability to date rock materials and to know where to look for the oldest rocks on Earth. Although the principles of such dating have more or less been known about since the discovery of radioactivity at the end of the 19th century, the techniques for doing so took several decades to develop. It was only in the wake of the Manhattan Project to develop the atomic bomb that radiometric dating began to be reliable and the age of the Earth was finally "pinned down" to around 4,550 million years.

Early universe

This 1,000-million-year-old universe, with stars being born, is based on deep field observations by the Hubble Space telescope. Because light has taken billions of years to reach Earth from these distances, these observations look back to the early history of the universe.

1200 MA The first multi-celled organisms date from the middle of the Proterozoic period

610 MA The first large marine animals appear

PHANEROZOIC EON

| 2500 MA PROTEROZOIC | 545 MA PALEOZOIC | 248 MA MESOZOIC | 65 MA CENOZOIC |

3900 MA Bombardment destroys early atmosphere and oceans

3800 MA first evidence of chemical life on the planet

Carboniferous sample) to 1,640 million years (a Precambrian sample). Within a year, he was able to propose the first geological time scale based on radiometric dating. Holmes's initial estimates of Earth's eras have held up remarkably well over time: for example, he placed the beginning of the Cambrian period at around 600 million years ago; today, 545 million years is the time frame that is largely accepted. In the early 1900s, however, Holmes's results appeared to be at odds with other methods in common use, and they were not met with immediate acceptance from all quarters.

Since Holmes's initial work in 1911, many improvements have been made to the process of radiometric dating. Of major importance was the discovery in 1913 that the atoms of a chemical element can exist in two or more different forms, called "isotopes", which are the same chemically, but have different atomic masses. Some isotopes are stable, and it is only the unstable ones that undergo radioactive decay. Of crucial importance for radiometric dating purposes, different unstable isotopes of the same element often have very different half-lives. For example, most uranium consists of the isotope U238, with a half-life of 4,500 million years; however, about 0.7 per cent is made up of the isotope U235, which has a half-life of 713 million years. Lead comes in several stable isotopic forms, some which occur in minerals only through the decay of other, radioactive isotopes, and yet others that were present in the minerals from when they first formed. It became apparent to Holmes and others that, for better accuracy, it would be necessary to measure isotope ratios (eg U238 to its decay product, the lead isotope Pb206), rather than just uranium to lead.

The discovery of isotopes initially complicated the process of radiometric dating, but did in time make it more precise. One initial effect was a reappraisal of Holmes's time scale. As a result of not compensating for then unknown factors, his computed ages were too high. Researchers realized that radiometric methods held promise for reassessing the Earth's age. In 1921, American astronomer Henry Russell (1877–1957) obtained 4,000 million years as a rough approximation of the age of the Earth's crust. This was based on an average of its maximum age calculated from its total uranium and lead content, and a minimum age based on the oldest known (at that time) Precambrian minerals. Over the following years, several more different ages for the Earth's crust were computed and published. These included 3,400 million years (Rutherford, 1929) and 4600 million years (Meyer, 1937).

Meanwhile, older and older rocks were being found in different parts of the world. By the 1940s, it became apparent that, to calculate an accurate age for the Earth, one piece of data was still needed – the ratio of different isotopes of lead in the Earth's crust at the time of its formation. Eventually, in 1953, the American geochemist Claire C. Patterson (1922–95) was able to infer this ratio through measurements on minerals of the Canyon Diablo meteorite, which contains very little uranium.

Meteorites, bits of mineral material orbiting the Sun that have only recently collided with Earth, are thought to have undergone very little change or reworking since their formation. As a result, their radiometric ages should be very close to the age of the solar system – and, by extension, the age of the Earth. Patterson was able to calculate an age for the meteorite of 4,550 million years, give or take about 70 million years. From studying the mixture of lead isotopes in the meteorite, and the same mixture in Earth crustal rocks, Patterson was also able to say that Earth was formed at the same time as the meteorite. Despite various further refinements to radiometric dating since then, his measure of 4,550 million years for the age of the Earth is still regarded as accurate.

During World War II, intense research on the atomic bomb led to major improvements in equipment for identifying and analysing isotopes. It became possible to detect minute quantities of specific isotopes and to measure their abundance with high precision. This in turn has led to highly accurate dating methods.

Modern radiometric dating of rocks is based on a number of different isotope combinations. For dating rocks between one and 100 million years old, an isotope with a shorter half-life is used – potassium 40 to argon 40 (half-life 1,300 million years). The decay of rubidium 87 to strontium 87 (half-life 47,000 million years) is another combination occasionally used.

Radiometric dating could originally only be applied to igneous rock, ie rock formed from the crystallization

of minerals from a molten material (magma) as it cooled. Indeed, until only recently, sedimentary rock was not suitable for radiometric dating. This is because the age of a specific grain in a sedimentary rock, such as a sandstone, is the age at which the mineral formed in its original igneous setting and not when it was locked into the sedimentary rock deposit. To discover the dates of sedimentary rock strata, geologists have traditionally had to find igneous rocks the age of which could be related to layers in the sedimentary strata (through superposition or crosscutting relationships). For instance, dating zircon from sediment illustrates certain problems with radiometric dating. A robust mineral, zircon may

have survived several cycles of erosion and many millions of years before being included in the sediment, making dating using the usual radiometric method inaccurate. A new radiometric method, however, is able to date tiny crystals that grow on zircon grains after deposition. As a result, it is now possible to date some sedimentary rocks where this type of additional mineral growth occurs. However, carbonate minerals which crystallize from carbonate-enriched water in some limestones, especially those deposited within caves, also contain radioisotopes which can be dated. They have proved particularly useful in dating climate changes during the Quaternary ice ages.

Oldest rocks

Early Precambrian (Hadean age) rocks at Isua in Greenland are among the oldest rocks still preserved at the Earth's surface. They were originally formed as volcanic deposits and sediments, mostly deposited in water around 3,800–3,700 million years ago. They have since been metamorphosed but still retain hydrocarbon residues of organic origin, showing that life was present in the waters they were formed in.

from them provoked more questions than answers. The Moon rocks were slightly younger than the oldest Earth rocks, but their chemical (isotope) composition was similar apart from a deficiency of iron. These findings eliminated two theories which, up to then, had been leading contenders for how the Earth–Moon system formed. The two bodies could not have formed together out of the same material – if they had, why would Moon rocks be younger than Earth rocks, and why would the Moon lack iron? And the Moon could not be a stray body that was captured by Earth's gravity. After all, why would it otherwise be so chemically similar to Earth?

In the 1980s, a new theory emerged for the origin of the Moon that seemed to fit the facts. Today, it is accepted as the most likely explanation. According to this theory, about 4,500–4520 million years ago – soon after it first formed and after differentiation into core and mantle – the Earth collided with another object about the size of Mars. As a result, a huge amount of crustal material was flung out and went into orbit about the Earth. Eventually, this material accreted to form the Moon.

Computer simulations have demonstrated that the collision theory is feasible. Such simulations have indicated temperatures of more than 10,000°K immediately after the event. As the ejected material accreted to form the Moon, an early crust formed from rocks called anorthosites (these now formed the lunar highlands), with the large basins known today as *mares* ("oceans") being excavated by impacts, then infilled with molten rock. Evidence from the well-preserved lunar surface indicates that the number of impacts lessened significantly as the solar system stabilized around 4,000 million years ago. The Earth would also have suffered the intensity of this late bombardment, but evidence of this is not preserved because of the dynamic reworking of the Earth's crust by plate tectonics.

If the collision did happen, in some ways it was a fortunate catastrophe because, without it, life on Earth may never have taken a hold. Among other consequences, the Moon eventually stabilized the Earth's rotation – if there were no Moon, the Earth would "wobble" more in its rotation, and we would have much more extreme seasons and weather.

As mentioned above, the oldest materials discovered so far with an identifiable Earth origin are some tiny crystals of a very tough mineral called zircon (zirconium silicate). These were found in 1999 within some sandstone rocks in Australia. They have been radiometrically dated as 4,400 million years old, implying they were formed a mere 150 million years after Earth itself.

The existence of these zircons has some deep implications for Earth's early history. Zircons most commonly crystallize in granite, a type of rock associated with continental crust, and there is every reason to suppose that this is how the Australian zircons also originally formed. Chemical analysis of the zircons using a technique called an ion microprobe has also shown that they contain an isotope of oxygen, oxygen–18, suggesting that they must have been formed in the presence of water.

Overall, the investigation of the zircons indicate that Earth may have developed continents and had some surface water – perhaps even oceans – as long as 4,400 million years ago. This is much earlier than previously supposed; however, they would have been destroyed by the late bombardment and had to re-form.

The zircons formed as crystals within molten granite that was cooling to form solid rock. The zircon-laden granite was eventually thrust upwards to form mountains, which later eroded. The granite vanished, but the zircons ultimately came to rest 3,000 million years ago in sandy Australian stream sediments. These sediments later hardened into rocks that subsequently were altered by heat and pressure.

If there were water on the Earth's surface 4,400 million years ago, the question arises – where did it come from? One possibility is that it was outgassed by volcanoes, but another favoured theory is that water was brought to Earth primarily by comets. Evidence from the lunar surface indicates that the Moon was subjected to intense bombardment from its formation until 3,900 million years ago. During that time, it was ravaged by more than a million major impacts from comets and asteroids. But Earth suffered bombardment, too, which may explain why no rocks survive from the first 500 million years of Earth's history – any crust that formed was soon destroyed by impacts. Also, as roughly half the content of comets is water, cometary impacts could certainly have brought plenty of water to the surface.

The "oceans from comets" theory has suffered some setbacks in recent years from studies of the gas emissions of comets. These indicate that the water in some comets, such as Comet Hale-Bopp, differs significantly from the water on Earth in its content of hydrogen isotopes. The water in other comets, however, such as Comet Linear (which broke up in 2000 as it passed the Sun), is very similar to Earth's. At present, the consensus is that comets may have contributed to Earth's oceans, but were probably not the sole source.

The composition of the Earth's atmosphere during its first few hundred million years of existence is almost entirely the subject of guesswork – there is very little evidence of what it may have contained, least of all in rocks. Some planetary scientists have speculated that the composition of the planet Jupiter today may be representative of the ancient atmospheres of the smaller planets. They base this conjecture on the fact that, because of its size, Jupiter has retained all the light gas

molecules it ever had, and these gases would have been present throughout most of the solar nebula.

On this basis, Earth's original atmosphere may have consisted mostly of hydrogen with some helium, but most of this is likely to have dissipated very quickly into space as a result of heating from the Sun. Earth's gravity would not have been strong enough to retain these light gases. The remaining atmosphere is likely to have contained some slightly heavier molecules formed from the most common elements – likely candidates being carbon dioxide (CO_2), methane (CH_4), ammonia (NH_3), nitrogen (N_2), water (H_2O), and some sulphur gases, but no free oxygen. Some of these gases were probably taken up in large part by dissolving in the oceans once these had formed, and others may have gradually been destroyed by light-mediated reactions. By around 3,500 million years ago, the atmosphere may have consisted mainly of nitrogen and carbon dioxide. Once photosynthesizing organisms appeared in the oceans and then on land, the stage was set for take-up of the carbon dioxide and liberation of free oxygen.

Star formation
The formation of a star such as our Sun begins when part of a nebula begins to coalesce into denser aggregations of gas. As the gas ball shrinks it becomes hotter, with temperatures and pressures at its centre becoming high enough to spark off nuclear reactions that convert hydrogen to helium and generate enough energy to make the star shine.

9

The Earth's future

Over recent decades, geological discoveries have shown that the Earth is a much riskier place to inhabit than previously thought. There have been both internally driven processes and external events that occur infrequently but on a much larger scale than anything before imagined. Plate tectonic movements have pushed continents from one hemisphere to another. Major impact events from space have repeatedly wiped out more than 50 per cent of all living organisms. Ice ages and vast outpourings of flood basalts have occurred several times. All these have had major impacts on the life of the past.

The good news is that these events have very low frequencies of occurrence and, despite all these vicissitudes, life has not only survived but has bounced back after the catastrophes. However, life is never quite the same after such an event.

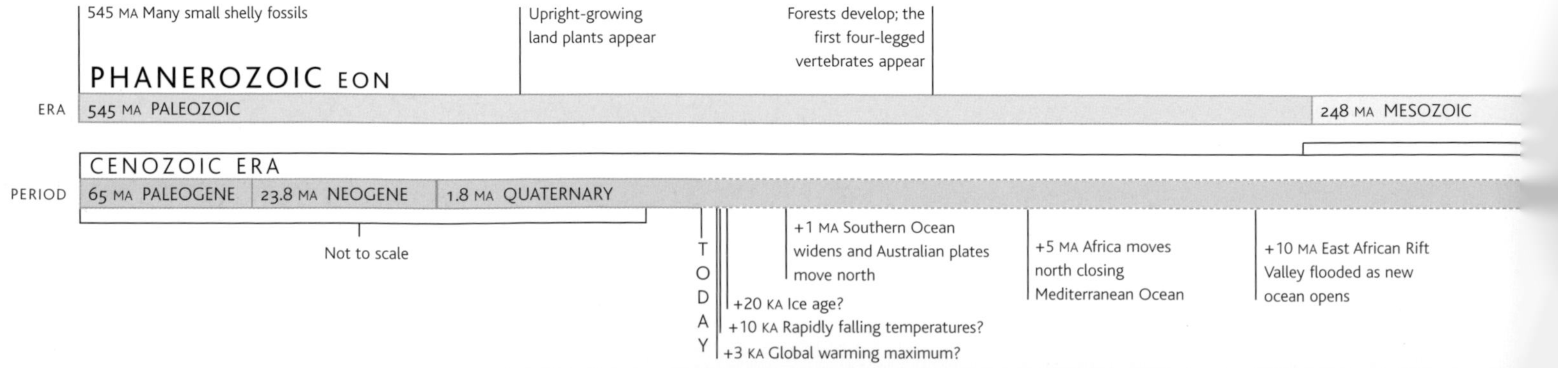

	545 MA Many small shelly fossils	Upright-growing land plants appear	Forests develop; the first four-legged vertebrates appear	
	PHANEROZOIC EON			
ERA	545 MA PALEOZOIC			248 MA MESOZOIC

	CENOZOIC ERA			
PERIOD	65 MA PALEOGENE	23.8 MA NEOGENE	1.8 MA QUATERNARY	

Not to scale

T O D A Y

+1 MA Southern Ocean widens and Australian plates move north

+20 KA Ice age?

+10 KA Rapidly falling temperatures?

+3 KA Global warming maximum?

+5 MA Africa moves north closing Mediterranean Ocean

+10 MA East African Rift Valley flooded as new ocean opens

As a result of these catastrophic events, old established groups such as the trilobites, the clubmosses (lycopsids), the dinosaurs, and pterosaurs are eclipsed and new groups, such as the crabs, flowering plants, songbirds, rodents, and humans come into dominance.

The earth sciences have a great deal to say about the Earth's future. From the study of these processes and events of the geological past, their mechanisms and range of magnitude and frequency, it is possible to develop scales of probability for future occurrences in much the same as an actuary calculates life expectancy based on family history, life style, and environment in which the subject lives.

While Earth as a planet has a very long way to go before any signs of "mortality" set in, a large number of highly significant events that will impact upon life will happen, some dramatically sudden, others noticeable over a lifetime, and others still more gradual but that will nevertheless have a very serious affect on life on Earth.

Blue planet

Two-thirds of the Earth's surface is covered with ocean water, while the remaining third is land. Vast areas of white cloud, made of water droplets, further demonstrate the vital role that water plays in the formation of our atmosphere, which protects the land and its inhabitants from harmful solar radiation.

| The first dinosaurs and early mammals appear | Birds and flowering plants appear | Primates and songbirds appear | 7 MA First hominids appear on Earth | TODAY |

65 MA CENOZOIC

| +20 MA Widening of Atlantic Ocean | | +40 MA Australia crosses the Equator; Antarctica moves north | +50 MA Major impact and extinction event? |

A dangerous world

**Kobe earthquake,
Japan 1995**
Recent geological investigation
has shown why some regions
of the Earth are more subject
to earthquakes and volcanoes,
but accurate prediction is still
some way off.

The abundance of living organisms on Earth, including
6,000 million or so humans, would seem to suggest that,
despite life's hazards, our planet is a relatively benign
place to live. Despite the hazards of reproduction, being
born as a relatively defenceless juvenile, the difficulty of
obtaining enough food to survive, and avoiding or

surviving disease and pestilence, organisms from algae to
zebras thrive on planet Earth. Certainly, compared with
the other planets in our solar system, Earth is hospitable.
The main reason why this is so, however, is a
combination of astronomical accidents.

Earth is neither too big nor too small, nor is it too
near or too far from the Sun. Its orbit, axis of rotation,
and rate of rotation all provide a narrow window that
allows the development of a protective atmosphere,
hydrosphere, and climate. This climate is generally not
too extreme and is thus conducive to life ranging from
the microscopic to the blue whale, a 30m (100ft) giant
which is the biggest animal ever to have lived.

Furthermore, vast amounts of heat energy are
required to keep our cosy little planet going. This energy
is transmitted from deep within the Earth and
transformed through some rather violent natural
processes such as volcanisms. The distribution in time
and space of the heat flow is uneven, with the result that
some parts of the Earth's surface are a lot more
geologically active than others. Fortunately, much of the
action takes place on the ocean floor, where new ocean-
floor rocks are generated, and in subduction zones, where
old ocean floor is returned back to the depths. Still, a lot
of the action does impinge upon life on land.

Earthquakes and volcanic eruptions are familiar
enough phenomena, even though they are still none too
predictable and the pressures of economics and
population growth result in very large numbers of people
living in regions that are known to be hazardous, such as
Mexico City, Naples, and San Francisco. Anyone looking
at an earthquake distribution map might perhaps think
twice about going on holiday to the Greek islands, parts
of Japan, Turkey, or California. And yet we still do so,
largely because the frequency of events that are truly
catastrophic is low within a decadal scale. So we take the
risk, just as we risk travelling by car or taking part in
hazardous sports. In the United States, the average risk of
death to an individual over a 50-year period is 1 in 100
from an auto accident, 1 in 5,000 from electrocution, 1

PRESENT + 10MA	+50MA	+100 MA
Major climate change	Major release eruption	Major impact

in 20,000 from a plane crash, 1 in 25,000 from a hurricane, 1 in 130,000 from lightning, and 1 in 200,000 from an earthquake. If you live in particular earthquake- or hurricane-prone regions, however, then the risks are higher – the figures are averaged out for the whole country. Global risks of death from a volcanic eruption are 1 in 30,000, but this figure is not very meaningful because it entirely depends on where you live, whereas many people now run the risk of suffering an auto accident.

Recent decades of geological investigation have uncovered a number of unexpected hazards for us to worry about. Fortunately, many of these have very low frequency rates, so low that they have not occurred within historic times. But they have happened and will happen again – it is just a matter of when. The fossil record tells us that, on a number of occasions in the past (at intervals of between 100 and 50 million years), life has been drastically cut back (as much as 70 per cent of all species) by extinction events. As far as we are able to tell, there are a number of different reasons for this. They range from a major impact event, such as that which occurred 65 million years ago at the end of Cretaceous times, to combinations of climatic and environmental change such as seem to have occurred at the end of Permian times 248 million years ago.

All the planets in the solar system suffer such impact events from extraterrestrial bodies over time. The most impressive of recent times was the one which hit Jupiter in 1994. Caroline and Eugene Shoemaker and their colleague David Levy spotted a string of 21 fragments approaching the planet in March 1993, at the Mount Palomar Observatory near San Diego, California. Known as Comet Shoemaker-Levy 9 (SL9), the fragments were estimated to be around a kilometre in diameter, and their estimated date of arrival on Jupiter was mid-July 1994. The world was waiting and was not disappointed: the impacts were visible from Earth even with small astronomical telescopes. Just one of the fragments (G), travelling at 60km (37 miles) per second, created an explosion as big as the K/T event. It was equivalent to some 100 million megatons of TNT and produced incredible thermal radiation in the stratosphere that would have fried anything living on the surface. The same would have applied to a similar event on Earth.

The global risk of death from an asteroid impact is 1 in 20,000, which in some senses is a more useful statistic than those for earthquakes for example, as the next asteroid hit could be anywhere on Earth (unlike earthquakes). Still there is no need to panic unduly. The statistic is based on the present total human population and the fact that although 500m- (1,640ft-) wide impacts occur every 10,000 years or so, 1km (³/₅ mile) asteroid hits are much rarer (every 100,000 years or so). However, the latter are likely to kill many hundreds of millions of people when they do impact.

Regrowth

Following Mount St Helen's most recent eruption in 1991 eruption soils have begun to reform on the mineral-rich volcanic debris and plants have begun to re-establish themselves. This volcano in the northwest of the United States is just one of many dangerous volcanoes which lie in a ring around the Pacific Ocean.

graptolites a group of marine colonial organisms, related to the living pterobranch hemichordates. Most are extinct and lived within branched organic skeletal tubes. Late Cambrian–Carboniferous with just a few surviving forms.

gymnosperms A large group of plants including ferns and conifers whose seeds are commonly held in cones or other modified shoots and whose ovules are not totally encased by tissue. Late Devonian–extant.

igneous rock Rock formed by the crystallization of minerals from a cooling molten material (magma).

invertebrates All those animals that lack a backbone.

isostasy The condition of equilibrium whereby the Earth's crust is buoyantly supported by the semi-plastic solid rock material of the mantle.

isotope A particular atom of an element that has the same number of electrons and protons as the other atoms of the element but a different number of neutrons.

kettle hole A depression in glacial outwash deposits, formed by the melting of a separated mass of buried glacial ice and the collapse of the overlying sediment.

(K/T) boundary Cretaceous/Tertiary boundary.

lissamphibians The large group to which all living amphibians belong (with over 4,000 species) and separate from extinct amphibian groups. Early Triassic –extant.

magma High-temperature molten igneous rock.

magnetite A strongly and naturally magnetic iron oxide mineral.

mantle A thick solid layer of rock within the Earth extending from below the crust to the core.

marsupials A group of primitive mammals that give birth to small immature young carried and suckled in a pouch on the mother's belly. Early Cretaceous–extant.

metamorphism The processes that produce structural and mineralogical changes in any type of rock in response to physical and chemical conditions differing from those under which the rocks originally formed.

monotremes A group of primitive mammals that lay large yolky eggs. Mid Cretaceous–extant.

moraine Rock debris that has been carried and deposited by a glacier.

notochord A cartilaginous dorsal stiffening rod that supports the body of all embryonic and some adult chordate animals. Precursor to the vertebral column.

obsidian Fine grained and rapidly cooled volcanic glass, usually black but sometimes brown or red.

orogenic belt A linear or arcuate mountainous zone in the Earth's crust, characterized by deformed and metamorphosed rocks, frequently associated with large deep plutonic intrusions and surface volcanoes.

peneplain A hypothetical surface, normally close to sea level, to which landscapes are reduced through prolonged mass wasting, stream erosion and sheet wash.

permafrost Soil or subsoil that is permanently frozen.

petrification or **permineralization** The preservation of hard parts of many organisms (both plant and animal) by mineral-bearing solutions after burial in sediment.

phylum (s), **phyla** (pl) Principal taxonomic category that ranks above class and below kingdom.

placentals A group of mammals that develop a special tissue, the placenta, by which embryos exchange nutrients and waste products with their mothers until they are born. Mid Cretaceous–extant.

placer deposit Superficial sedimentary deposit laid down by water, containing economic quantities of valuable minerals.

planetismal A body of rock and/or ice formed in the primordial solar nebula, from which larger planetary bodies are thought to have formed by coalescence.

plate tectonics A synthesis of geological and geophysical observations in which the plates of the Earth's outer rigid lithosphere move relative to each other, diverge to form new oceans, converge to form volcanic island arcs and mountain ranges and slip past one another as transform faults.

pluton Any major intrusive body of igneous rock formed deep within the crust by partial melting and then slow cooling of magma.

precocial Those young hatched or born in an advanced state and able to feed themselves almost immediately.

prokaryotes Those primitive micro-organisms in which the genetic material is not contained within a discrete membrane-bounded nucleus, but scattered throughout the cell.

reptiliomorphs A group of extinct and primitive reptile-like tetrapods. Mid Carboniferous–early Triassic.

RNA Ribonucleic acid, a nucleic acid found in all living cells.

roche moutonnée A glacially shaped mound on a glaciated bedrock surface.

rostroconchs A group of primitive extinct bivalved molluscs. Earliest Cambrian–end Permian.

rudists A group of extinct bivalve molluscs, many of which had large, conical, coral-shaped shells and formed reef-like structures. Cretaceous.

scaphopods A small group of marine univalved molluscs (about 1,000 species) with hollow curved tusk-like shell, open at both ends. Middle Ordovician – extant.

sedimentary rock A rock formed by the burial and consolidation of sediment settled out of water, ice or air and accumulated on the Earth's surface under the influence of gravity, either on dry land or under water.

speciation The evolution of different species.

stomata The leaf pores plants use to regulate their "breathing" (gas exchange).

stratum (s), **strata** (pl) A layer or series of layers of sedimentary rock generally separated by bedding planes which approximate to the horizontal when the sediment was originally deposited but may since have been inclined or folded by earth movements.

stromatolites Laminated and mound-shaped calcareous sedimentary structures produced by alternating microbial films and thin layers of sediment.

subduction The movement of one crustal plate, generally the denser, under another, with the descending plate sinking into the mantle and eventually being consumed at depth.

taxon (s), **taxa** (pl) One of a hierarchical group of organisms ranging from species to kingdoms.

Tertiary period Original term for life's third age, now divided into two periods – Neogene and Paleogene.

till Generally non stratified sedimentary rock debris deposited directly by glacial ice.

tillite A sedimentary rock formed by the compaction and cementation of till.

trilobites An extinct group of marine arthropods characterized by a segmented body divided longitudinally into three lobes and transversely into three sections from head to tail. Early Cambrian–end Triassic.

uniformitarianism The principle originating with James Hutton (1726–97), stating that the laws of nature now prevailing have always prevailed and that, accordingly, the results of processes now active resemble the results of like processes of the past.

varanids A group of lizards (monitor lizards). Late Cretaceous–extant.

vertebrates The large grouping which includes all animals with a backbone or spinal column ranging from fish to mammals. Late Cambrian–extant.

bibliography and useful websites

Benton, Michael J *Vertebrate Palaeontology* (Chapman & Hall 1991)

Benton, Michael J *When Life Nearly Died: The Greatest Mass Extinction of All Time* (Thames & Hudson, 2003)

Briggs, Derek EG and Peter R Crowther *Palaeobiology II* (Blackwell Science, 2001)

Clarkson, ENK *Invertebrate Palaeontology and Evolution*, (Blackwell Science, 1998)

Condie, Kent C *Plate Tectonics and Crustal Evolution* (Butterworth-Heinemann, 1998)

Currie, Philip J and Kevin Padian (eds) *Encyclopedia of Dinosaurs* (Academic Press, 1997)

Hancock, Paul L and Brian J Skinner (eds) *The Oxford Companion to The Earth* (Oxford University Press, 2000)

Kingdon, Jonathon *Lowly Origin: Where, When, and Why Our Ancestors First Stood Up*, (Princeton 2003)

Knoll, Andrew H *Life on a Young Planet: the First Three Billion Years of Evolution on Earth* (Princeton, 2003)

Lewin, Roger *Principles of Human Evolution* (Blackwell Science, 1998)

Lewis, CLE and SJ Knell *The Age of the Earth: from 4004 BC to AD 2002* (Geological Society, 2001)

Palmer, Douglas *The Atlas of The Prehistoric World* (Marshall Editions, 2000)

Palmer, Douglas *Fossil Revolution: The Finds that Changed Our View of the Past*, (HarperCollins, 2003)

Press, Frank and Siever, Raymond *Understanding Earth* (WH Freeman & Co 2000)

Van Andel, TH *New Views on an Old Planet: a History of Global Change* (Cambridge University Press, 1994)

Wilson, RCL, Drury SA, and Chapman JL *The Great Ice Age: Climate Change and Life* (Routledge 2000)

The World Wide Web is an excellent source of information about geology. However, as with so much web "data", you have to be careful about the reliability of the information. Generally, those sites set up by major national institutions such as geological surveys, museums, universities and publishers of internationally creditable journals can be relied upon, as can many of the links they provide because they are not trying to sell you anything. I have found the following useful:

University of California Museum of Paleontology, Berkeley www.ucmp.berkeley.edu; **Natural History Museum, London** www.nhm.ac.uk; **The Geological Society, London** www.geolsoc.org.uk; **The Palaeontological Association, UK** www.palass.org; **American Museum of Natural History** www.amnh.org; **British Geological Survey** www.bgs.ac.uk; **Geological Society of America** www.geosociety.org; **National Museum of Natural History, Smithsonian Institution, Washington D.C.** www.nmnh.si.edu/paleo/links.html; **United States Geological Survey;** www.usgs.gov/index.html; *National Geographic* www.nationalgeographic.com; *American Scientist* www.amsci.org/amsci; *Discover* www.discover.com; *Nature* www.nature.com; *Scientific American* www.sciam.com; *Science* www.sciencemag.org

index

acknowledgements

(in page order) 2 Senckenberg, Messel Research Department, Frankfurt am Main; 5 Science Photo Library/Joe Tucciarone; 9 Ministère de la Culture, A Chéné - Centre Camille Jullian - Centre National de Préhistoire, Périgueux; 11 Getty Images/Image Bank/Jeff Hunter; 12 Still Pictures/Jacques Jangoux; 13 NASA; 14 Still Pictures/M & C Denis-Huot; 15 Still Pictures/Fred Bruemmer; 17 Science Photo Library/William/Ervin; 18 top left NASA; 18 top right NASA; 18 bottom Ministère de la Culture, A. Chéné - Centre Camille Jullian - Centre National de Préhistoire, Périgueux; 19 Tony Waltham Geophotos; 21 NASA GSFC Visualization Analysis Laboratory; 23 Tony Waltham Geophotos; 24 Lochman Transparencies/Alex Steffe; 25 AKG-Images/Antiquarium Museum, Pompeii; 26 Tony Waltham Geophotos; 27 Science Photo Library/George Bernard; 29 Douglas Palmer/Dr Phil Lane; 30 Douglas Palmer; 31 Agence France Presse/Ravanelli/Getty Images; 32 Douglas Palmer; 34 Getty Images/Taxi/Harvey Lloyd; 35 M.O. & J Plassard; 36 Geology Institute, Neuchâtel University, Switzerland; 37 Douglas Palmer; 39 Illustrated London News Picture Library; 40 Science Photo Library/Philippe Plailly/Eurelios; 41 Douglas Palmer; 43 The Society of Antiquaries of London; 44 Science Photo Library/Volker Steger/Nordstar - 4 Million Years of Man; 45 Science Photo Library/Tom McHugh/Field museum, Chicago; 46 Tony Waltham Geophotos; 47 Science Photo Library/Dr Juerg Alean; 49 Senckenberg, Messel Research Department, Frankfurt am Main; 50 © The Royal Society, London; 53 © The Natural History Museum, London; 55 Science Photo Library/John Reader; 56 Oxford Scientific Films/Clive Bromhall; 57 Science Photo Library/John Reader; 58 Senckenberg, Messel Research Department, Frankfurt am Main ; 59 Senckenberg, Messel Research Department, Frankfurt am Main; 60 Hessisches Landesmuseum Darmstadt; 62 NASA/Jacques Descloitres, MODIS Land Science Team; 63 top Science Photo Library/Dr Ken Macdonald; 63 bottom Science Photo Library/Dr Ken Macdonald; 64 Science Photo Library/NASA/JPL; 65 NASA/Jacques Descloitres, MODIS Land Rapid Response Team, NASA/GSFC; 67 Science Photo Library/Chris Butler; 69 Science Photo Library/Jim Amos; 71 © The Natural History Museum, London; 73 Science Photo Library/Sheila Terry; 74 Tony Waltham Geophotos; 75 Tony Waltham Geophotos; 77 top Pratt Museum of Natural History at Amherst College; 77 bottom Pratt Museum of Natural History at Amherst College; 78 Science Photo Library/Joe Tucciarone; 79 Museum des Sciences Naturelles de Belgique; 81 © The Natural History Museum, London; 82 © The Natural History Museum, London /M Long; 83 Dr Thomas Martin; 84 Dr David Dilcher; 85 top right Carnegie Museum of Natural History, Pittsburgh/Mark A Klingler; 85 bottom left © The Natural History Museum, London; 87 Professor Michael Hambrey; 89 Science Photo Library/Professor Walter Alvarez; 90 Peter Sheldon; 93 Science Photo Library/Joe Tucciarone; 94 Professor Greg Retallack; 95 Science Photo Library/Dr David Kring; 96 Laszlo Keszthelyi; 97 Science Photo Library; 98 Professor Daniel Schrag; 99 Tony Waltham Geophotos; 101 Douglas Palmer; 105 Science Photo Library/Ted Clutter; 106 Professor Michael Hambrey; 107 Peter Sheldon; 108 Professor Michael Hambrey; 111 The Art Archive/Musée National d'Art Moderne, Paris/Dagli Orti; 112 Science Photo Library/John Durham; 113 Professor Andrew Scott; 114 left Science Photo Library/Worldsat International Inc; 114 right Science Photo Library/Worldsat Productions/NRSC; 119 Reproduced by permission of the British Geological Survey. © NERC. All rights reserved. IPR/41-7C; 120 Dr Jennifer A Clack; 121 Dr Jennifer A Clack; 123 Science Photo Library/Sinclair Stammers; 124 Science Photo Library/Sinclair Stammers; 125 Reproduced with the permission of the Minister of Public Works and Government Services Canada, 2003 and Courtesy of Natural Resources Canada, Geological Survey of Canada; 126 Degan Shu, Northwest University, Xi'an and Simon Conway Morris, University of Cambridge; 129 Professor Michael Hambrey; 130 Reproduced by permission of the British Geological Survey. © NERC. All rights reserved. IPR/41-7C; 131 NASA/GSFC/LaRC/JPL, MISR Team; 132 Reproduced with the permission of the Minister of Public Works and Government Services Canada, 2003 and Courtesy of Natural Resources Canada, Geological Survey of Canada, photo: S McCracken; 133 Lochman Transparencies/Len Stewart; 134 © H J Hofmann; 135 University of California Museum of Palaeontology, Berkeley; 136 David M Rudkin, Royal Ontario Museum; 137 Dr T P Crimes; 138 Professor Michael Hambrey; 141 Tony Waltham Geophotos;142 Corbis/Bettmann; 143 top Science Photo Library/Dr Kari Lounatmaa;143 bottom Tony Waltham Geophotos; 144 Science Photo Library/NASA; 145 NASA; 147 Professor Shuhai Xiao and Professor Andrew H Knoll; 149 Science Photo Library/Space Telescope Institute/NASA; 150 Getty Images/Hulton Archive; 151 Mary Evans Picture Library; 152 Professor Sam Bowring; 155 Corbis/James L Amos; 156 NASA JSC/Carl Allen; 157 top left NASA; 157 top right NASA; 157 bottom left NASA; 157 bottom right NASA; 159 Science Photo Library/R Ermakoff/Eurelios; 161 Science Photo Library/NASA; 162 AKG-Images/Rijksmuseum, Amsterdam; 166 Tony Waltham Geophotos; 167 Tony Waltham Geophotos; 168 Still Pictures/Lo Tsung Hsien/UNEP; 169 Tony Waltham Geophotos